PREVENTIVE CONSERVATION - FROM CLIMATE AND DAMAGE MONITORING TO A SYSTEMIC AND INTEGRATED APPROACH

Reflections on Cultural Heritage Theories and Practices

A series by the Raymond Lemaire International Centre for Conservation

ISSN: 2576-3075
eISSN: 2576-3083

Volume 5

PROCEEDINGS OF THE INTERNATIONAL WTA - PRECOM³OS SYMPOSIUM, LEUVEN, BELGIUM, APRIL 3–5, 2019

Preventive Conservation - From Climate and Damage Monitoring to a Systemic and Integrated Approach

Edited by

Aziliz Vandesande

Raymond Lemaire International Centre for Conservation, Department of Civil Engineering, KU Leuven, Heverlee, Belgium

Els Verstrynge

Department of Civil Engineering, KU Leuven, Heverlee, Belgium

Koen Van Balen

Raymond Lemaire International Centre for Conservation, Department of Civil Engineering, KU Leuven, Heverlee, Belgium

CRC Press
Taylor & Francis Group
Boca Raton   London   New York

CRC Press is an imprint of the
Taylor & Francis Group, an **informa** business

A BALKEMA BOOK

Cover photo: Monumentenwacht in action (Province East Flanders, Belgium), © Monumentenwacht

'Reflections on Cultural Heritage Theories and Practices' book series

3. Innovative Built Heritage Models
    Edited by Koenraad van Balen & Aziliz Vandesande
    ISBN 978-1-138-49861-7 (HB)
    ISBN 978-1-351-01479-3 (eBook)

4. Professionalism in the Built Heritage Sector
    Edited by Koen Van Balen & Aziliz Vandesande
    ISBN 978-0-367-02763-6 (HB)
    ISBN 978-0-429-39791-2 (eBook)

5. Preventive Conservation – From Climate and Damage Monitoring to a Systemic and Integrated Approach
    Edited by Aziliz Vandesande, Els Verstrynge & Koen Van Balen
    ISBN 978-0-367-43548-6 (HB)
    ISBN 978-1-003-00404-2 (eBook)

CRC Press/Balkema is an imprint of the Taylor & Francis Group, an informa business

© 2020 Taylor & Francis Group, London, UK

Typeset by Integra Software Services Pvt. Ltd., Pondicherry, India

Library of Congress Cataloging-in-Publication Data

LC record available at https://lccn.loc.gov/2020028366
LC ebook record available at https://lccn.loc.gov/2020028367

Applied for

Published by: CRC Press/Balkema
            Schipholweg 107C, 2316XC Leiden, The Netherlands
            e-mail: Pub.NL@taylorandfrancis.com
            www.crcpress.com – www.taylorandfrancis.com

ISBN: 978-0-367-43548-6 (Hbk)
ISBN: 978-1-003-00404-2 (eBook)
DOI: 10.1201/9781003004042
https://doi.org/10.1201/9781003004042

Table of contents

Committees   vii

Introduction

Preventive conservation - from climate and damage monitoring to a systemic approach   3
A. Vandesande & E. Verstrynge

From knowledge background to practical applications

A coevolutionary approach as the theoretical foundation of planned conservation of built
cultural heritage   11
S. Della Torre

Innovation and diversification of brick, Susudel – Ecuador   19
G. Barsallo, T. Rodas, V. Caldas, F. Cardoso, C. Peñaherrera & P. Tenesaca

Monitoring of China's built heritage since 1950s: Historical overview and reassessment of
preventive conservation   27
W. Meiping, H. Shi & L. Xinjian

A brief review on preventive conservation and its application in China's conservation
background   37
Q. Rong, B. Liu & J. Zhang

Preventive and Planned Conservation: potentialities and criticalities, strategy and tools,
lessons learned   47
R. Moioli

Monumentenwacht model and new initiatives

Preventive conservation model applied in Slovakia to monitor built heritage damage   59
P. Ižvolt

The Traditional Buildings Health Check: A new approach to the built heritage in Scotland   67
S. Linskaill

Quality of restoration of monuments: The role of Monumentenwacht   71
S. Naldini, G. van de Varst, S. de Koning & E. van de Grijp

Preventive and planned conservation for built heritage. Applied research in the University of
Porto   77
T.C. Ferreira

Preventive monitoring and study of insect damage of carpenter bees to timber components of
chinese historic buildings   87
Y. Gao, Y. Chen, D. Xu, E. Li, J. Li, Z. Ge & Y. Zhou

Condition assessment and monitoring in Milan Cathedral: Putting risk assessment into practice 93
L. Cantini, F. Canali, A. Konsta & S. Della Torre

The role of the university in maintaining vernacular heritage buildings in the southern region of Ecuador 103
G. García, A. Tenze & C. Achig

Damage diagnosis and monitoring of case studies

MDCS - a system for damage identification and monitoring 113
R.P.J. van Hees & S. Naldini

Monitoring of water contents and temperatures of historical walls with interior insulation in Switzerland 119
C. Geyer, B. Wehle & A. Müller

Immediate measures to prevent further damage to the wall frescos of the "Ritterhaus Bubikon" 125
K. Ghazi Wakili, Th. Stahl, D. Tracht & A. Barthel

Energy retrofit of historic timber-frame buildings – hygrothermal monitoring of building fabric 129
C.J. Whitman, O. Prizeman, J. Gwilliam, P. Walker & A. Shea

3D Laser scanning for FEM-based deformation analysis of a reconstructed masonry vault 137
A. Drougkas, E. Verstrynge, M. Bassier & M. Vergauwen

Contribution of photogrammetry and sensor networks to the energy diagnosis of occupied historical buildings 145
S. Dubois, J. Desarnaud, Y. Vanhellemont, M. de Bouw, D. Stiernon & S. Trachte

Author index 153

Committees

ORGANISING COMMITTEE

Prof. Els Verstrynge, *University of Leuven, Belgium*
Dr. Aziliz Vandesande, *University of Leuven, Belgium*

INTERNATIONAL SCIENTIFIC COMMITTEE

Kris Brosens, *Triconsult NV, Belgium*
Ali Davis, *Historic Environment Scotland, Scotland*
Anastasios Drougkas, *University of Leuven, Belgium*
Luigi Barazzetti, *Politecnico di Milano, Italy*
Sebastiaan Godts, *Royal Institute for Cultural Heritage (KIK-IRPA), Belgium*
Cui Jinze, *University of Leuven, Belgium*
Gabriela Lee Alardín, *Universidad Iberoamericana, Mexico*
Silvia Naldini, *TU Delft, The Netherlands*
Teresa Patricio, *ICOMOS - Board member, Belgium*
Wido Quist, *TU Delft, The Netherlands*
Eduardo Rojas, *University of Pennsylvania, United States*
Maria Eugenia Siguencia Avila, *University of Cuenca, Ecuador*
Koen Van Balen, *University of Leuven, Belgium*
Bert van Bommel, *Rijksvastgoedbedrijf, The Netherlands*
Dionys Van Gemert, *University of Leuven, Belgium*
Rob van Hees, *TU Delft, The Netherlands*
Yves Vanhellemont, *WTCB, Belgium*
Michiel Van Hunen, *Rijksdienst voor Cultureel Erfgoed, The Netherlands*
Birgit van Laar, *Monumentenwacht Vlaanderen, Belgium*
Nathalie Vernimme, *Onroerend Erfgoed, Belgium*
Luc Verpoest, *University of Leuven, Belgium*

Introduction

Preventive conservation - from climate and damage monitoring to a systemic approach

A. Vandesande
Raymond Lemaire International Centre for Conservation, Department of Civil Engineering, KU Leuven, Belgium

E. Verstrynge
Building Materials and Building Technology Section, Department of Civil Engineering, KU Leuven, Belgium

INTRODUCTION

This volume reports on the lectures presented during the international WTA-PRECOM³OS conference on preventive conservation, jointly organized by WTA-Nederland-Vlaanderen, the Raymond Lemaire International Centre for Conservation and the Civil Engineering Department of KU Leuven (Leuven, Belgium, 3-5 April, 2019), which brought together 30 speakers and 165 participants from 12 countries.

WTA is an international organisation committed to the encouragement of scientific research and its practical application in the field of building maintenance and monuments preservation. WTA international has regional groups in Germany, Swiss, Czech Republic, and The Netherlands-Flanders. The latter group, WTA-NL-VL joined its spring symposium with the international WTA days to support the organization of the international WTA-PRECOM³OS Conference. The contributions to the symposium in Dutch were published on the website WTA-NL-VL.org.

In addition, this volume highlights the 10th anniversary of the UNESCO Chair on Preventive Conservation, Monitoring and Maintenance of Monuments and Sites (PRECOM³OS), which was established at the Raymond Lemaire International Centre for Conservation in 2009, in collaboration with Monumentenwacht Vlaanderen and the University of Cuenca in Ecuador. Over the years PRECOM³OS driven built heritage research, training and collaborations have grown into a large international network including the including – but not limited to – the Universidad de Oriente in Santiago de Cuba, the Polytechnic University of Milan and Southeast University in Nanjing.

1 CONTENT AND SCOPE OF THIS VOLUME

1.1 *Preventive conservation*

Preventive conservation aims at implementing minimum interventions, the least destructive of all interventions which inevitably occur in built heritage conservation, through a continuous process of identifying, assessing, analysing and monitoring expected damages, possible risks and the overall state of conservation of built heritage structures in combination with planned interventions based on a cyclical methodological model which takes into account non-linear patterns of deterioration and aims as much as possible to avoid incompatible and postponed interventions which lead to reactive treatment patterns, unforeseen detrimental damage and additional resource investments. Herewith, defining and aligning of different interventions is based on a global understanding of the building, a long-term perspective and sufficient information and experience to translate the interventions in practice through a qualitative implementation. This preventive conservation process is based on a combination of condition-based and scheduled building maintenance, albeit with the difference that the continuous process maintains a dynamic approach towards available resources, the buildings' performance and fluctuating external factors. Moreover, preventive conservation differentiates from building maintenance as it recognises that standard approaches and strategies to understand and maintain existing structures do not align with the specificities of heritage values, historic interventions and additions to a building, local building technologies and cultural context (Vandesande 2018).

1.2 *Practice-based research*

Since the mid-1990s, preventive conservation in the built heritage sector became increasingly endorsed within international and European policies, which in turn was also reflected in different EU-funded research projects and by international organisations that started implementing a pro-active approach towards monitoring and maintenance.

Ongoing research in context of the PRECOM³OS UNESCO Chair demonstrates that preventive conservation in the built heritage sector contribute to cost-effectiveness, the improved protection of heritage values, the reduced risk for accumulating deterioration and additional damage, the prolongation of the physical service life of buildings and building parts and the empowerment of local communities in dealing with heritage (Vandesande et.al. 2018). Although a pro-active approach towards preventive approaches and monitoring gained importance in the research field since the mid-1990s, as 'a reflection of the growing commitment to improving management frameworks for care of cultural heritage through the use of monitoring, which is understood as a key component of the management process' (Stovel 2008: 15), the built heritage sector has largely failed to put these findings in long-term sustainable practices.

Therefore, this volume consists of three main chapters that deal with practice-oriented and practice-based research. The first chapter is entitled *From knowledge background to practical applications*, which contains articles which consolidate existing local and international research on preventive conservation. These theoretical findings are either combined with a retrospective analysis of preventive conservation case studies or translated into practical innovations.

The second chapter, entitled *Monumentenwacht model and new initiatives* gives a central role to Monumentenwacht as a true operational legacy of preventive conservation efforts. For the first time, the practice and experiences of most existing Monumentenwacht organisations and initiatives are brought together in one publication. This consolidated information offers a valuable and substantive contribution that highlights preventive conservation and monitoring, with reflections of future improvements and innovations.

The final chapter in this volume is entitled *Damage diagnosis and monitoring of case studies* and deals with case-study based research, in which preventive conservation is implemented in practice through specific measurement techniques and monitoring tools. Several contributions focus on climate monitoring and control as a major part of a preventive conservation approach. In addition, the possibilities of visualisation techniques for supporting on-site inspection and diagnosis are highlighted.

2 FROM KNOWLEDGE BACKGROUND TO PRACTICAL APPLICATIONS

The keynote of this chapter is a contribution by *S. Della Torre*, who has developed some of the basic concepts of preventive conservation which are still used today. In this article, the author aims to enhance the theoretical foundation of the conservation process, involving continuous care, a long-term vision and openness to evolving values. A co-evolutionary approach is introduced, to rethink the relationships between conservation and time and enable the shift from restoration as an event to conservation as a process. This approach is illustrated by the case study of Como, Church of Saints Cosma and Damiano in Italy. One practical and innovative example of this co-evolution approach is presented by *G. Barsallo et al.* The authors introduce the Later-Eris research-based project in South-Ecuador, which deals with innovation and diversification of the brick. The objective of the project was to diversify traditional *panelón* brick production in the village of Susudel which is currently threatened by mono-production, over-supply and unsustainable prices. The research-based project entails a multidisciplinary approach with chemical, physical and mechanical material analysis as well as research by design on traditional local patterns. The project outcome contributes to the continuation of brick production in the Susudel area and guarantees that traditional vernacular buildings can be maintained in the future based on compatible building materials.

While these two papers introduce the concept of co-evolution, the next articles in this chapter contribute the existing state of the art on preventive conservation research and concepts. *W. Meiping, H. Shi, L. Xinjian* provide a retrospective overview of built heritage monitoring in China since the 1950s, based on four case studies including two World Heritage Sites and two nationally listed buildings. The retrospective analysis aims to assess how effective the monitoring of built heritage in the context of preventive conservation was carried out. Following this retrospective overview, the next article provides a more future-oriented approach towards preventive conservation in China. *Q. Rong, B. Liu, J. Zhang* provide a review on preventive conservation and its application in China based on working package modules tailored for the local conservation sector. These working packages cover key phases in the preventive conservation process such as regular inspection and maintenance, risk assessment, monitoring, vulnerability index evaluation, but also more practical components such as health condition investigation through panoramic aerial photography, 3D laser scanning and non-destructive testing technology.

In the final paper of this chapter *R. Moioli* provides, from a professional point of view, a critical reflection on Preventive and Planned Conservation practices in the Lombardy Region in Italy, where since 2003 several field tests took place on historic buildings. The analysis conducted by the author focusses on the evolution of preventive conservation activities and their impact related to building owner, legal frameworks and competencies of professionals. Finally, the author indicates that a viable first step towards tackling these challenges is to set-up local centres based on the monumentenwacht experiences, which is further elaborated in the next chapter.

3 MONUMENTENWACHT MODEL AND NEW INITIATIVES

The Monumentenwacht model is an operational, efficient and cost-efficient built heritage monitoring system on the historic urban environment level. Their central rationale is timely identification and correction of defects on historic structures to reduce deterioration of the fabric and prevent major consequential damage. In practice, their core activity is supporting private and public built heritage owner-managers in the preservation of their properties through a periodic monitoring system.

The Netherlands and Flanders play a pioneering role as a country and region where Monumentenwacht is successfully realised, respectively since 1973 and since 1991. Since the Council of Europe campaign 'Europe, a common heritage' in 2000, monumentenwachten are increasingly known in various European countries and regions. Other organisations made interesting attempts to set-up similar models in their own respective countries, i.a. the UK (Maintain Our Heritage in Bath), Denmark (Bygningsbevaring), Germany (Denkmalwacht), Hungary (Műemlékőr). More recently there are new initiatives with interesting additions to and successful adaptions of the original monumentenwacht model, i.e. Pro-Monumenta in Slovakia, the Traditional Buildings Health Check Scheme in Scotland and Fixus in Lithuania.

The first article in this chapter is by *P. Izvolt*, who introduces the Pro-Monumenta model in Slovakia, founded by the Ministry of culture of the Slovak Republic and Monuments Board of the Slovak Republic. The article provides an overview of the current state of conservation of protected monuments in Slovakia, the organisational scheme of Pro-Monumenta, the most frequently occurring defects on the built heritage, the main re-occurring problems in the organisation and future development perspectives.

The next article in this chapter is by *S. Linskaill*, who provides an insight on the Traditional Buildings Health Check, a new approach to built heritage in Scotland based on the monumentenwacht model. This Scottish scheme was created in 2013, when the Historic Environment Scotland and the Construction Industry Training Board (Scotland) funded a 5-year pilot project to practically address the large backlog in repairs of traditional domestic buildings in Scotland. The article delivers and overview of how the medium and long-term objectives at the beginning of the pilot were implemented over a period of 7 years. The pilot project clearly demonstrates that there is a valid and verified role for the Traditional Buildings Health Check service in providing impartial expert advice to building owners to ensure that they can make timely and informed decisions. In the future, the intention of the monumentenwacht based model is to stimulate demand for a construction sector with the skills and capacity to appropriately maintain and repair Scotland's built heritage.

Next, *S. Naldini* analyses the contribution of Monumentenwacht in The Netherlands to enhance the quality of interventions on historic buildings. Based on case study analysis, the potential and limits of their practice is discussed. The author finds that a key role of Monumentenwacht is to make private owners aware of the value of their monuments, as well as the respective building materials and techniques. In addition, the author introduces the project 'Monumentenwacht moves', which was developed in co-operation with Delft University of Technology, and which aims at creating a uniformity in reporting and monitoring among the various stakeholders and institutions working in the field.

Next to these articles on monumentenwacht based experiences, which are here uniquely published together for the 10th PRECOM³OS anniversary, this chapter also contains contributions that reinvent the model into for very specific needs and contexts. *T. C. Ferreira*, presents research on preventive and planned conservation strategies applied on the specific case study of the Faculty of Architecture of the University of Porto. The preventive conservation methodologic framework designed for the case study combines different actions focused on the life cycle of the building, such as bibliographic and archive research, fieldwork and data collection on materials and techniques, inspection and monitoring with NDT (thermohyigrometer, thermography, decay mapping, visual inspection), drawing of constructive details, consultation of contractors for maintenance planning and the design of infographic illustrated manuals for users. All this data is compiled in a digital database and App, specifically developed for built heritage.

The next two papers in the chapter each provide a specific approach to the monitoring component of a preventive conservation approach. Firstly, *Y. Gao et al.*, introduce preventive monitoring of insect damage by carpenter bees to timber components of Chinese historic timber buildings. The research team combined the field monitoring and laboratory monitoring to conduct a full-year documentation and study on the different types of infestation on wooden building components. The research resulted in a large amount of basic data, which provided the researchers with a baseline for the preventive conservation of the insect damage of carpenter bees to wooden components of Chinese historic buildings. The monitoring approach in their research thus not only contributes to establishing a preventive approach, but also contributed to fill the gap in China's domestic research on the impact of carpenter bees on timber buildings. The second article introduces a new analytical tool which puts the theory of risk assessment into practice. Herewith, *L. Cantini et al.* focus on a single case study, the Milan Cathedral, due to the availability of a large body of knowledge and expertise by the managing body of the cathedral and their close collaboration with the Politecnico di Milano. Based on the analysis of different risk and condition assessment approaches, the authors present a new procedure for detecting risk conditions through a set of indicators. The aim of this procedure is to define an index connected to the frequency of the inspection activities necessary for guaranteeing safety aspects concerning the building and the people attending its spaces. While this tool holds a large potential to identify future cracks and failure risks, the first simulations indicated some difficulties in managing the risk evaluation procedure for large monumental elements and limits for more detailed decay pathologies. The research team continues to further develop the analytical tool and combine it with a 3D digital model of the cathedral to refine the quality of the data.

Finally, *G. García, A. Tenze and C. Achig*, present a preventive conservation approach developed and led by the University of Cuenca in Ecuador which focusses on modest examples of vernacular earth-based architecture. The innovative component of this approach lies in the intrinsic link with community participation in both urban and rural areas of the country. The variety of activities developed by the University of Cuenca in order to enhance the conservation of traditional vernacular architecture has resulted in various observations. The authors define the need for a new methodological framework, the creation of promotor groups beyond the academic space and the activation of follow-up committees.

4 DAMAGE DIAGNOSIS AND MONITORING OF CASE STUDIES

The keynote paper by **R. van Hees and S. Naldini** stresses the importance of a thorough investigation before starting interventions on historic buildings, and consequently introduces an system for damage identification and monitoring. This Monument Diagnosis and Conservation System is and online tool which supports the diagnosis of damages based on damage atlases, which encourage a clear communication among users as this forms a necessary basis for monitoring.

In contrast to this paper on a very systemic damage assessment tool, the other articles in this chapter can be situated more on the level of site specific case studies, whereby the focus of each article is on specific diagnosis and monitoring approaches, from in-situ monitoring to hygro-thermal analysis and detailed geometric surveys. Several contributions focus on hygro-thermal condition monitoring and control as a major part of a preventive conservation approach.

C. Geyer et al. focus on risks involved in energetic refurbishments of walls in historical buildings due to high moisture contents in the construction. Therefore, the post-isolation moisture content in the layers of the construction is predicted numerically. Predicted values of the simulation program reproduce measured values of moisture content and temperature within a small error margin, which can be used as a safety margin for future predictive calculations and risk mitigation in the design phase of an energetic refurbishment. Also *C. J. Whitman et al.* focus on predicting risks involved in energetic

retrofit. Yet, their research is dealing with historic timber-framed buildings. In situ monitoring highlights that in some instances, the combination of incompatible materials, flawed detailing, poor workmanship and lack of controlled ventilation can facilitate biological attack. They developed hygro-thermal simulation models to investigate the influence of orientation, climatic conditions and in-fill material on the walls' hygro-thermal behaviour. Although no prolonged periods of conditions favourable to biological decay were identified, and these initial results are supported by monitoring of test panels under laboratory conditions, further long term monitoring is required to understand, prevent and mitigate decay that can be found on-site in historic timber-framed buildings.

While the previous papers target the retrofit design phase, the contribution by *K. Ghazi Wakili et al.* focusses on urgent measures that halted ongoing moisture deterioration and prevent future damage to the frescoes of the entrance hall to the unheated Ritterhaus chapel. By closing all openings and installing a controlled ventilation system, condensation on the walls has been avoided. This was made possible by a monitoring-supported and model-based operating of the ventilation, to blow in the outdoor air only when its absolute humidity was lower than that of the indoor air during the critical period.

Further contributions make use of advanced digitisation techniques, such as 3D laser scanning and photogrammetry, to establish digital models of the heritage structure for diagnosis support. As a first example, *A. Drougkas et al.* demonstrate the need for accurate geometric survey of historical vault structures in order to acquired deformation data to validate numerical modelling approaches. Focusing on a complex reconstruction project involving a masonry vault at the Royal Academy of Fine Arts in Ghent, practical aspects of damage monitoring, geometric survey and computational analysis of historic structures are jointly presented and addressed. As the vault was reconstructed in lime-based mortar, 3D monitoring during removal of the supporting centering provided essential data for validation of a numerical model for deformation predictions, and proved that accurate representation of the 3D geometry is vital for predicting deflections of complex vaulted structures. And finally, **S. Dubois et al.** show how photogrammetry and wireless sensor networks can be combined to produce rich datasets for energy diagnosis of historical buildings, while keeping disturbances for occupants at a minimum. A case study is presented to illustrate the deployment of the proposed methodology. Only two site visits allowed to capture a large quantity of descriptive and performance information. The proposed diagnostic methodology shows clear benefits in terms of efficiency for the building energy diagnosis, especially when dynamic energy models are implemented, to explore retrofitting scenarios and support the decision making.

5 REOCCURRING OBSERVATIONS

The central observation is the importance of flexibility in implementing preventive conservation approaches to different built heritage sites, which are characterised by a variety of local conditions, climate, physical environment, design of buildings and used materials. The findings of many of the papers suggest that setting-up new preventive conservation initiatives will need to be flexible in how they can implement and make use of existing monitoring and maintenance methods and tools. Too rigid rules and regulations, implementation guidelines or project-based plans might counteract possibilities to change built heritage conservation practices. Nevertheless, there is a need for a new methodological framework that considers key challenges and solutions, enabling systemic decision making among different actors in the long-term preventive conservation process.

A second important observation that can be directly derived from the findings of this volume is the need to resist using only new technologies in an effort to simplify complex project contexts. Many of the case studies presented discuss how the implementation and use of technologies can be an incentive for stakeholders as they can simplify the preventive conservation process (from damage diagnosis to monitoring and scheduling maintenance) to an extent that it enables stakeholders in making reasonable decisions both within and outside their of specialization. Again, rules and regulations, implementation guidelines or project-based plans should be reluctant towards simplification. Instead, they should embrace the complexity and interdisciplinarity of the preventive conservation process to support creative planning to allow interaction and the maximization of the benefits for all stakeholders, end-users and owners. In line with this, there is need to find balance between the effectiveness of preventive conservation and digital technologies. The meaningful combination of on-site monitoring, laboratory testing and changes throughout building history can provide a wide array of stakeholders with the often much needed baseline for setting-up preventive conservation plans for specific building topologies. The effort of integrating existing interdisciplinary data, or the mere start of collecting data on built heritage sites, is often time consuming without immediate utility.

6 FUTURE RESEARCH DIRECTIONS

A number of issues are still open and need to be addressed by future research. Most of the authors in this volume speak of preventive conservation models as something that is readily available on any project and that just needs to be tailored so that it can be meaningfully used. Making such an assumption is reasonable to understand the larger organizational and policy implications of implementing preventive conservation. However, it does not allow one to understand in detail how monitoring and maintenance can be used within the larger management process or designing interventions, where information is only slowly generated, often iterated and changed, and only frozen at very specific points within the process. One future research direction is to learn how to handle such fluid information, which could draw on design-thinking theories. We would expect such studies to provide important insights into how stakeholders can not only handle the fluid preventive conservation implementation processes but also the momentary nature of information and data collection throughout this process.

Another closely related issue that still needs to be addressed by future research efforts is the kinds of information and types of data that are required by different actors at various stages within the preventive conservation processes. Most of presented research assumes that a ready-made information model for preventive conservation is simply available, but largely remains silent about how these information models are structured and what information they contain. Future research is required to provide detail about the ontological content of the multitude of information that is required in such a model and which models already exist (apart from the Monument Diagnosis and Conservation System presented by van Hees and Naldini) an which models from building and infrastructure projects can serve as a basis. Given the rapid digital transformation of the sector, a systemic information model for preventive conservation that can represent and abstract its long-term and cyclical process in a meaningful way will be important in the years to come, so that stakeholders and actors can use them as support for decision-making. Together these studies could finally lead to a better understanding about what information owners, managers and governments require to make good policy decisions about the built environment.

In addition, there is a need for more systematic monitoring of intervention performance against quality requirements through the preventive conservation life cycle. High quality interventions contribute to local communities' well-being, but examples of low-quality interventions give rise to complaints from experts and citizens and may even damage irreplaceable historical elements and their environment. Shared performance indicators and protocols for quality of interventions in the preventive conservation process can support the shift from descriptive project evaluations to more quantified, scientific and transparent approaches. The development of such performance indicators and protocols requires a platform that brings together the research community, public and private actors and policy makers at local, regional, national and international levels concerned with the impact assessment and quality of interventions on the built environment and heritage sites.

REFERENCES

Maintain our Heritage. 2004. *Historic building mainten-ance – A pilot Inspection service*, Bath (Maintain our Heritage).

Stovel, H. 2008. 'Origins and Influence of the Nara Docu-ment on Authenticity'. *APT Bulletin*, 39 (2/3): 9–17.

Vandesande A., Van Balen K. (2018). "Preventive conser-vation applied to built heritage: A working definition and influencing factors". In: Van Balen K., Vandesande A. (Eds.), *Innovative Built Heritage Models*. (Raymond Lemaire International Centre for Conservation series, 3). CRC Press/Balkema (Taylor & Francis group): 63–72.

Vandesande A., Van Balen K., Della Torre S., Cardoso F. (2018). "Preventive and planned conservation as a new management approach for built heritage: from a physical health check to empowering communities and activating (lost) traditions for local sustainable development". *Journal of Cultural Heritage Manage-ment and Sustainable Development*, 8 (2), 78–81. doi: 10.1108/JCHMSD-05-2018-076.

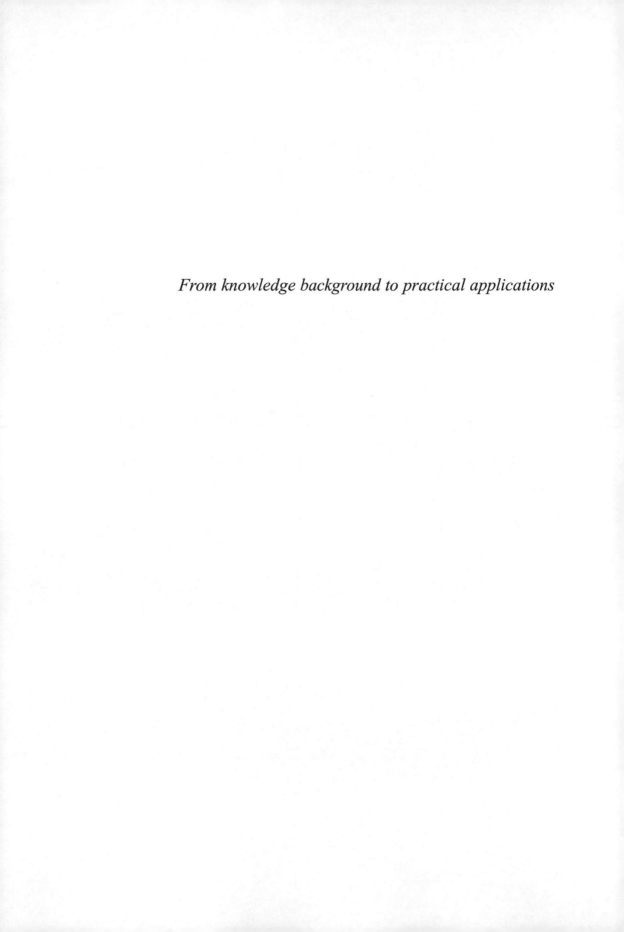

From knowledge background to practical applications

Preventive Conservation - From Climate and Damage Monitoring to a Systemic and Integrated Approach – Vandesande, Verstrynge & Van Balen (eds)
© 2020 Taylor & Francis Group, London, ISBN 978-0-367-43548-6

A coevolutionary approach as the theoretical foundation of planned conservation of built cultural heritage

S. Della Torre

Department of Architecture, Built Environment and Construction Engineering, Politecnico di Milano, Milan, Italy

ABSTRACT: The metaphor of coevolution is applied to conservation of built cultural heritage, in order to enhance the theoretical foundation of conservation process, involving continuous care, long-term vision and openness to evolving values. Implementing the concept has an impact on preservation practices, including both the ones typically meant as preventive, i.e. inspection and monitoring, and the ones usually understood ouside the preventive sphere such as reuse.

1 INTRODUCTION

Even if consisting of several research streams developed on different premises and in different cultural contexts, it is by now possible to acknowledge the existence of one movement for the preventive conservation of cultural heritage, thanks to reflections and debates matured through almost two decades.

The preservation of heritage can be carried on through different strategies and tools, which can be "curative", as they address damages once they have been detected, or "preventive", as the strategy aims at addressing the causes of loss and controlling continuously the transformation processes. To explain the differences between the two approaches, previous publications deepened the exam of the well-known analogy of preservation with medicine, ending up into an opposition similar to "Curative (medical) Care" versus "Public Health" (Pracchi et al. 2010; Van Balen, 2015).

In the past, the practiced preservation strategies were definitely oriented towards a "curative" approach, often identified in the restoration of the historic and aesthetic values. Maintenance was quoted in all the Charters, from the Charter of Athens (1931) to the Venice Charter (1964) and also in the Burra charter, but the vision of preventive conservation as a system is a recent development, produced through researches carried out for almost two decades. Reviewing the output of these years, the recurrence emerges of some themes, which are characteristics of the subject: the importance of regular maintenance, the making of inspections carried out following the Monumentenwacht model, the information management, the engagement of owners and users (Van Balen & Vandesande 2013). The common label that unifies all the researches on preventive conservation may be identified in the reference to time and long-term vision, which are the same

references of developments like cost-effective conservation, or sustainable conservation. The implementation of sustainability paradigms on heritage gave an important contribution to strengthen the theories and innovating the practices (among many others: Cassar 2009; Pereira Roders and Van Oers 2011; Biscontin and Driussi 2014). In order to go one step further, some references to epistemology and evolutionary thinking can be proposed, in order to build theoretical foundations. In this paper the metaphor of coevolution is analysed as inspiring for the purpose, trying to describe conservation process as a coevolutionary one. Conservation process is seen in a perspective, which tries to be holistic. The impact on practices is then discussed in detail, focusing on reuse, inspection, assessment and risk management, documentation, financial evaluations.

2 CONSERVATION AS A COEVOLUTION PROCESS

2.1 *The metaphor of coevolution*

The conservation of a building cannot be simplified taking into account only few of the many dimensions of its complexity: any innovation would fail, if implemented without the awareness of the systemic nexus. All the relevant issues have to be considered, such as traditional craftsmanship and new technologies, the engagement of communities, the compatibility of uses, the management of resources, and the economic convenience of timely intervention.

Together with the systemic approach, the discussion on time is a key factor to renovate the scientific discourse on all the above listed topics. The mental attitude of people thinking in preventive terms is not concentrated on the link of the present to the past, but the future time is considered, in terms of aims of the actions as well as in terms of probability of the events.

In order to get a deeper understanding of this mindset, the use of a metaphor can be useful. This is justified because metaphors are not mere figures of speech, but they constitute an important mechanism of the mental processes (Martinez et. al. 2001: 965; Haggis 2004: 182). A mighty metaphor that helps to elaborate these issues is Coevolution. It comes from biology sciences, ant it is related to Darwinian processes affecting species, which have strong interactions with each other, so that the process can be described as a reciprocal evolutionary change.

Coevolutionary models have been widely applied as metaphors, that is outside the biology field.

The implementation in history and archaeology to explain trends and changes was quite obvious, as it was a way to enrich the understanding of evolution processes, in which the Darwinian thinking could be an obvious reference (Ames 1996; Mesoudi, Whiten, Laland 2006).

The implementation in economics has been more similar to the problem we are dealing with. Evolutionary Economics aims at understanding changes as the result of events that sometimes are definitely random, and forces, which are often antagonist to each other. However, some authors observed that while it is possible to describe some coevolution processes as definitely emergent from the complexity of the systems, in some other processes coevolution is somehow guided by external actions (Cuervo-Cazurra, Martin de Holan & Sanz 2014). It is worthy notice that Evolutionary Economy has been inspiring for modelling preventive conservation as a system, and its development as a relevant change for a local sustainable development (Vandesande 2017). In the field of Evolutionary Economics, the metaphor of coevolution has been explicitly invoked to explain trends involving environment and resources (e.g. Norgaard 1994; Kallis G. & Norgaard 2010), to investigate reasons for the competitive advantage of locations, and above all to develop models for innovation and change management (e.g. Van den Bergh & Stagl 2003).

Among the fields, in which the concept of coevolution has been successfully used as an interpretation tool, the cultural landscape sector is maybe the closest to historic preservation. In Italy, the "territorialist" approach used to treat the territory as a highly complex living system, developing sophisticated methods to deal with ecosystems (Magnaghi 2017). The definition of cultural landscapes as complex adaptive systems encompasses both the concepts of emergence and coevolution (Rescia et al. 2012). Implementing the concept of "extended evolution" (Laubichler & Renn 2015), Niles and Roth proposed to understand traditional agriculture landscapes as "living knowledge systems", to be preserved not as the relics of a time gone by, but as resources for development through the interaction with new actors and societal processes (Niles & Roth 2016).

2.2 *The existence of built heritage as a coevolution factor*

What we are proposing in the present paper is that built cultural heritage as well should be understood as a living knowledge system.

The proposed implementation of a coevolutionary approach aims at enlightening the potential influence that the presence of heritage produces on environment and society. Implementing the metaphor implies giving to coevolving objects a (metaphorical) condition of living beings, maybe absurd for showcased things, but no so strange for the complexity of an inhabited building, whose performance depends on socio-cultural processes. If, referring to the concept of "extended evolution", as introduced by Laubichler and Renn, Niles and Roth observe that preservation of cultural heritage looks simpler than it is the case for agricultural landscapes, we propose that built cultural heritage as well should be understood as a living knowledge system.

The concept of "living knowledge systems" could influence even the first step of any preservation process, that is the recognition in heritage objects of some cultural value. Recognition (of the object as a work of art) was introduced by Cesare Brandi as the beginning of the restoration process (Brandi 2005). The following reflections, taking into account also the contribution given by the late Paul Ricoeur (Ricoeur 2005), are widening this concept from the recognition of the artistic essence to the acknowledgment of any kind of values, as so many and diverse are the reasons why heritage may trigger attention and protection, or inspire innovation processes and creativity. Undoubtedly, approaching recognition under an evolutionary perspective opens several new options, such as the appreciation of imperfection, as the premise of future developments, or the understanding of peculiar cases, as evolutionary niches, which bear the witness of something disappearing and almost forgotten.

Therefore, evolutionary (and coevolutionary) thinking is consistent with the step to anthropological approaches that, few decades ago, produced a renovation of heritage understanding, stepping from the cult for works of art to the interest for the relationships, which build the territorial tissue (Montella 2003). In this perspective, the study on cultural heritage is no longer a matter of chasing masterpieces, but of identifying networks of significance. The theses on conservation developed in Italy after 1975, focused on making restoration less selective (Bellini 1999), elaborated reflections that go far beyond the Venice Charter theoretical background: an idea of conservation as a multi-phases process (including prevention and maintenance) had still to be introduced, but the foundations of a new attention to evolution, openness and complexity were laid.

This theoretical research was looking for new references, exploring the legacy of different authors, which had not yet been considered by the previous theory of restoration, focused on aesthetics: just Cesare Brandi had been inspired also by the philosophers of phenomenology. Referring to the cognitive step "from being to becoming" proposed by an author like Ilya Prigogine and, it has been proposed that conservation should focus on preserving above all the potentialities for coevolution (Della Torre 1999). In this step, what matters is the displacement of the focus from the level of the facts to the level of potentialities. As no longer limited to the most significant features, conservation takes the task to forward into the future also the witnesses of those memories, which sometimes are still to be recognized and investigated.

This means that beyond recognized values, there is a reservoir of knowledge that still has to be explored. To give an example, the history of construction aims at understanding how in the past, and in the frame of traditional techniques, some problems had been solved, or at least had been taken into account to make buildings more durable and resistant to external actions. In seismic countries, far before the modern standards, it is possible to recognize devices and solutions, which actually contribute to a good behaviour under earthquake actions. The recent experience of L'Aquila in 2009 entailed an intense activity of investigation and intervention on historic buildings, which in their long lifespan already faced strong seismic events and keep the signs of damages, reparations and devices studied in order to decrease their vulnerability, often on the basis of just an empirical thinking (e.g. see Indirli et al. 2013). Thus a world of forgotten knowledge was rediscovered, along with a number of hidden details that are not part of the recognized values, but became extremely significant now, and therefore cannot be given up, also because they can inspire a more proficient and sustainable way to retrofitting the buildings, merging structural and historic knowledge (Bartolomucci 2013).

The above described example leads to an issue with relevant practical implications, the concept of "living knowledge systems" can be the way to give a positive theoretical foundation to the discussion about traditional practices vs. new technologies, which should not be dealt with as a matter of ideologies, but as an amazing opportunity for a coevolution work. It is worthy to notice that the concept of historic buildings as living knowledge systems is fundamental to the debate on traditional techniques vs. new technologies, which should not be a comparison of ideologies, but a positive opportunity for the development of a coevolutionary work (Vandesande et al. 2018).

2.3 Coevolution and the understanding of the conservation process

A coevolutionary approach implies a change throughout the preservation agenda. In several researches the process of preservation activities has been analysed in order to identify the outputs, the actors, the aims of all the activities. A scheme of the process has been proposed (Della Torre 2018) in the frame of the CHANGES (Cultural heritage Activities: New Goals and benefits for Economy and Society) JPI Heritage+ project (Figure 1).

The scheme tries to breakdown the whole conservation process, in order to understand the relationships among the activities, which are identified as "preventive conservation" and the more traditional ones, under the hypothesis that all the activities contribute to conservation if they are correctly planned. In other words, remedial actions and reuse interventions can also be oriented to a long-term vision, and sometimes a preventive system gets launched as the follow-up of a traditional restoration. Therefore, the alternative between preventive activities and curative actions is not an ideological one, but it is the consequence of practical decision based on management control and risk assessment.

This means that what matters is the long-term vision, and historic buildings require a management system, which could be designed as a conservation plan dealing with the issues of facility management. On the other hand, the concept of preventive conservation has been developed, as said above, in the direction of people engagement and community involvement, which are important opportunities for the management of cultural properties: people and communities take responsibilities as users, owners, spectators, actually co-creating contents and values related to the experience of heritage and landscape.

The long-term conservation management should be thought as the management of change, according to a definition of conservation that has been given by both Sir Bernard Fielden and Amedeo Bellini (Fielden 2003: vi; Bellini 1999). This definition can sound a bit paradoxical: it has been bitterly rejected by prestigious Authors, who see in this openness to change a threat to "the classic values of conservation" (Petzet 2010: 9), but the threat of a conflict between conservation and development is much

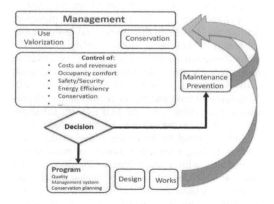

Figure 1. Layout of the process.

13

more dangerous. It is much more productive to think heritage as co-evolving with the context. According to the current approach, conservation is the defence against the changes induced by environment and society: in a coevolutionary approach, change is unavoidable, but the presence of heritage can influence the process, and therefore the aims and the criteria, opening to different options, otherwise unimaginable, and definitely powerful just for the activities that engage persons and communities.

2.4 *Coevolution and the management of change: Towards a holistic understanding of preventive conservation*

In the field of historic preservation, preventive conservation is an innovation, and its implementation is a change management challenge.

For two decades several researches have been carried on the implementation and planning of prevention and maintenance, as well as of a long-term vision in interventions. The progressive widening of the discourse in these studies is definitely evident, as the analyses went deeper in the investigation of issues and biases. Without spoiling the importance of maintenance practices, it has been realized that in all the fields, in which prevention is central, strategies are implemented with a systemic approach, not limited to standing alone actions.

The thought on preventive conservation had therefore to widen its horizon, particularly to the issues of management and societal concerns. Therefore, social and economic impacts have been studied under the planning perspective, reasoning no longer on single activities and phases of the conservation process, but in holistic and systemic terms (Van Balen & Vandesande 2018), developing the discourse on economic and societal impacts (CHCfE Consortium. 2015; Van Balen & Vandesande 2016) and on community involvement (Van Balen & Vandesande 2015). Besides the awareness that fundamental benefits are produced by heritage-related activities thanks to the connections they are capable to create, the reference to time emerged as the general key of the discourse. Namely the reference is to the long-term perspective, built cultural heritage and its long-lasting presence: planning takes place in this frame, and in this frame the benefits of prevention come to reality.

The above introduced oneness of the conservation process, entails a holistic perspective: on one hand, curative activities should tend to be thought with a long-term horizon and put in consistency with prevention and maintenance, as well as with participatory activities, on the other hand maintenance itself should be inspired by the awareness of evolutionary thinking. In fact, a reductive concept of maintenance is consistent with a static idea of significance and values, and it risks to turn into a mechanical process of remedial actions against decay, losing not only the control on the implications of prevention and maintenance, but also the real impact of routine operations on the uniqueness of historic-material contexts. If significance is dynamic and plays with memories and values, which require devoted attention to be detected and valorised, no direct operation on heritage could be reduced and simplified.

3 THE IMPACT ON PRACTICES

3.1 *Coevolutionary (more than adaptive) reuse*

Under the holistic perspective, planned conservation encompasses also post-intervention management. An architecture cannot be understood by sight only, but using and inhabiting it; use is necessary for valorisation and conservation, and most of the preservation effort concerns the compatibility and the reasonable changes required by the societal evolution. A recent paper deals specifically on the implementation of the coevolutionary metaphor in the issues concerning the reuse of historic premises (Della Torre 2019).

Changing the use is often a good solution for a better compatibility, as requirements are often evolving even for the same function, so that "the reinsertion of a new utility" (Rypkema 2012) could require less alterations than complying with the enhanced requirements for keeping the original or current function. Beyond physical conservation, nonmaterial significance is very important as well. Outside the conservation sector, moving from interior design, adaptive reuse has been recently proposed as an autonomous discipline (Plevoets & Van Cleempoel 2013), claiming an historical background that is the same of restoration (Viollet-le-Duc, Ruskin, Riegl, Boito…). In the reality, differences are not negligible: adaptive reuse is carried on in a short-term perspective with a "research by design" methodology, which encompasses the target of keeping and showing a nonmaterial significance: but significance is understood once forever and thought as a permanent character of the place itself (Plevoets & Prina 2017). The sense of place, in a coevolutionary perspective, is instead highly dynamic, as it emerges from the meeting of the object with the percipient subject.

Reuse projects should be carried on with the awareness that user experiences will often be unpredictable co-creation processes. Not surprisingly, marketing research on tourism industry devoted a wide literature on the management of co-creation processes engaging customers (e.g. see Prahalad & Ramaswamy 2004; Prebensen & Xie 2017; Buonincontri et al. 2017). Even if cultural heritage sites should never be reduced to tourism destinations, this means that open and co-operative interpretation of cultural contents is far more intriguing than the communication of simplified lessons.

For this reason, the will to take the synthetic expert's interpretation as the foundation of the restoration and reuse project looks very naïve. If the presentation of a reused facility should communicate the significance of the place as stated once forever by the expert, the visitor would be driven to miss a large deal of the knowledge hidden in the historical substance. Such an approach leads to simplify the historic complexity, reducing the diversity of the messages, which all together constitute the future dynamic significance of the place.

An example can show the complexity produced by historical reuse practices, and how many potential messages can be wasted if the expert's interpretation goes straight forward, implementing selective criteria and overlooking other meanings.

It's the case of a small Romanesque church, transformed into a dwelling house around 1670. A Renaissance fresco had been detached from the apse in 1898, then the building was not protected nor studied, and the detached fresco was conserved in the local museum as coming from a demolished church. In the 1980's the church was re-discovered, surveyed, published (Crippa 1984), then restored in 1988. The restoration was undoubtedly a cultivated action, leading to a number of interesting archaeological and artistic discoveries (Brogiolo & Zigrino 1993; Natale 1998; Caimi, 2001; Casati 2009-2010; Quattrini & Natale 2017), but the features introduced in the old reuse, that is splitting the internal volume in two storeys by the insertion of a floor and an illogical balcony in the middle of the Romanesque apse (Figure 2), were unthinkingly removed (Figure 3), as if they had no potential to tell other stories or to trigger the co-creation of a more comprehensive appreciation and a different future reuse: even if many other experts expressed their preferences for a definitely different and maybe more sustainable reuse project.

The case helps to explain the comparison of two different understandings of the significance, one more static as based on mainstream values, and one more dynamic, that is open to the emergence of new contents and values, exploiting the evolutionary potential of what is embedded in layered historic buildings.

Figure 3. Como, Church of Saints Cosma and Damiano. The apse during restoration works (1991).

A dynamic conception of the significance is important because in the reality the co-creation processes are unpredictable, as the society is diverse and evolving, so that no unilaterality is productive (Lagerkvist 2006). The unpredictable experiences developed by the encounters of the historic substance and the variety of the public, is just what coevolution means in terms of the nonmaterial inspiring value of historic places. In other words, coevolutionary approach confirms the step from the traditional expert-centred model, the expert being an historian or a designer, to a more user-centred, or community-centred model, based on the awareness that the significance is produced by a bit of co-creation, and maybe heritage can carry out its educational function only through the active engagement of the public (Silberman 2016). On the other hand, engagement of people is not only for enjoyment, but also for the activities encompassed in the conservation process. This way, the educational mission will be completely accomplished, involving the contents related to citizenship and responsibilities and social inclusion, on the line pointed out by the Faro Convention (Therond & Trigona 2009).

3.2 Inspection, condition assessment, risk management

A care-oriented approach deals with objects as living beings, carrying their own individual stories and destined to have trajectories through time, in which the

Figure 2. Como, Church of Saints Cosma and Damiano. The apse before restoration works (1984).

"Becoming" enhances their significance. The awareness of this cultural foundation, derived from the concept of coevolution, changes the operational attitudes, even in inspections and assessment.

The key step from the restoration as event to conservation meant as a continuous process has been also synthetized in the claim "from cure to care", but it also triggered the understanding of analysis, conservation and structural restoration as a cyclic process (Van Balen & Verstrynge 2016). It is worthy to underline that, in this cycle, the anamnesis phase has a lot to do with the exploration of the knowledge inbuilt in historic fabric: understanding the construction, the structural and seismic history, the hidden resources entail exactly an approach to the historic structure as a living knowledge system, which can be strengthened better if its intimate logic has been investigated and implemented.

Through the dialogue with preventive conservation as developed in the museum field (Staniforth 2013), the topic of risk management has been transferred into the preventive conservation of built environment. A subtle difference can be underscored between this approach and the typical attitudes developed from the restoration practices. While the restorer tends to set up the remedial measures against what has already happened, the prevention specialist looks forward to guess what is next to happen, focusing on the causes. A popular didactic example, used to explain this difference, is the alternative between looking at the current state of conservation of the picture, or at the quality of the light produced by the lamp. It is a change of mental attitude, which already has been codified for museums, in terms of standards, procedures, required skills.

The step from cure to care involves the different relationships between the historic building and its context. It is worthy to remark that this relationship is no longer unidirectional and merely adaptive in coevolutionary perspective, because the presence of the historic object influences the cultural, social, and economic development of the context. This can happen through different mechanisms: heritage can produce an effect on feelings, can have an inspirational action, can give limits but also directions to spatial planning, and so on. In some cases, these effects are produced in a spontaneous way and sometimes inadvertently, in other cases the presence of heritage can become an opportunity to develop projects aimed at enhancing the resilience and sustainability of regions. That is why coevolution means much more than adaptation, as it includes a positive attitude.

The consequent proposal of substituting the risk assessment to the traditional synthetic evaluations of urgency has been accepted by the European Standard *Conservation process – Decision making, planning and implementation* (EN 16853:2017). We may argue that it has been a fundamental conceptual step from maintenance thought as an occasional small

restoration to prevention recognized as an autonomous managerial practice (Della Torre 2010).

In 2018 in Italy as well the Preventive Conservation movement had the opportunity to celebrate the ten years of the "Conservazione Programmata" calls issued by Fondazione Cariplo (Moioli & Baldioli 2018). Looking at the different editions of this initiative, which made many pilot projects possible involving an important number of technicians, owners, users and contractors, it is worthy notice that in the last years the emphasis has been moved from maintenance works to prevention, taking into account also big hazards, and raising awareness on the priority of prevention in a seismic country like the Italian peninsula.

3.3 *Documentation and knowledge management*

The oneness of the conservation (and valorization) process entails that the management of knowledge through the different activities is one of the specific features of planned conservation. To say better, it is what enables conservation to be preventive, and effective on the long run. Therefore, important advancements have been carried out in practices and methodologies concerning documentation, such as survey, monitoring and information systems. The development of digital technologies made possible and even easy the exchange and reuse of data, in ways that could not be imagined before. Interoperability is the key for further advancements and implementation of these opportunities (Della Torre & Pili 2019), which support the transition towards practices consistent with coevolutionary thinking.

3.4 *Financial evaluations*

Long-term thinking should entail a radical change in financial evaluations as well. Some researches on maintenance programs pointed out that owners are not willing to anticipate expenses for benefits, which are expected far in the future (Dann 2004). Even if the consensus about the convenience of preventive attitudes is unanimous on paper, described behaviors are generally un-responsive, as practitioners seem to be not interested in predictive cost analysis, and decision makers tend to underestimate big hazards, not investing enough in the prevention of floods or earthquakes. In terms of behavioral economy, such unresponsive attitudes are not surprising: they cannot be addressed by plain regulation (Alemanno & Sibony 2015), and both empowerment and nudging are long-term strategies, based on the concept of evolving interactions among the players.

4 CONCLUSIONS

Reasoning on Planned/Preventive Conservation and Coevolution enables not only to enhance change management practices, but also to understand better which is the object of the care. Beyond material authenticity and nonmaterial significance, which are both important, the ultimate target of preservation can be described as keeping the potentialities for coevolution. In other words, conservation means to keep those features, which are relevant to the values that the heritage object will be able to produce in the future, thanks to the emergence of new relationships and co-creation processes, in which the presence of the past will open to new opportunities while preserving identities and treasures.

If this is the target, it will be important to remark also that its description in these terms has been possible rethinking the relationships between conservation and time, and this leads to the necessity of consistent and planned policies. Coevolutionary thinking strengthens the idea that conservation process is one, and all activities become more productive as they are thought as interrelated.

The same theoretical foundation supports the idea of broadening the scope of preventive conservation to include, sharing "the responsibility for preservation to a larger fraction of society than traditional conservation practices do" (Van Balen, 2017: 715). The shift from restoration as an event to conservation as a process, and the shift from the expert-centered to the user-centered approach are the two sides of the same coin, and support each other.

ACKNOWLEDGEMENTS

The author wants to thank Koen Van Balen for the effort produced to develop the concept of Preventive Conservation and to preach it worldwide; The authors wants to affectionately remember Luigia Binda, who introduced him and Koen to each other, starting a long lasting coevolution.

REFERENCES

Alemanno, A. & Sibony, A. (eds) 2015. *Nudge and the Law: A European Perspective (Modern Studies in European Law)*. Oxford and Portland: Hart.

Ames, K. 1996. Archaeology, Style and the Theory of Coevolution. In H.D.G. Maschner (ed), *Darwinian Archaeologies*: 109–131. New York: Plenum Press.

Bartolomucci, C. 2013. Structure and architecture: the illogical results of considering them two separated entities, after the 2009 earthquake in L'Aquila. In Cruz (ed), *Structures and Architecture: Concepts, Applications and Challenges*: 1621–1628. London: Taylor & Francis.

Bellini, A. 1999. De la restauración a la conservación; de la estética a la ética. *Loggia: Arquitectura y restauración*, 9: 10–15.

Biscontin, G. & Driussi, G. (eds.) 2014. *Quale sostenibilità per il restauro?* Venice: Arcadia Ricerche.

Brogiolo, G.P. & Zigrino, L. 1993. La chiesa dei SS. Cosma e Damiano di Como. Analisi stratigrafica degli alzati. In Uboldi, M., *Carta Archeologica della Lombardia. Como. La citta` murata e la convalle*: 98–101. Modena: Panini.

Buonincontri, P., Morvillo, A., Okumus, F. & Van Niekerk, M. 2017. Managing the experience co-creation process in tourism destinations: Empirical findings from Naples, *Tourism Management*, 62: 264–277.

Brandi, C. 2005. *Theory of Restoration*. Florence: Nardini.

Caimi, R. 2001. La basilica dei SS. Cosma e Damiano: la situazione attuale e l'indagine archeologica In *Prime pietre. Gli esordi del Cristianesimo a Como: uomini, fonti e luoghi*: 129–140. Como: Iubilantes.

Casati, M.L. 2009-2010. L'altare rinascimentale della Chiesa dei Santi Cosma e Damiano di Como. Contributo alla ricomposizione del contesto originario. *Rivista Archeologica dell'Antica Provincia e Diocesi di Como* 191-192: 429–437.

Cassar, M. 2009. Sustainable heritage: challenges and strategies for the twenty-first century, *ATP Bulletin* 40/1: 3–11.

Crippa, M.A. 1984. La chiesetta dei SS. Cosma e Damiano. In *S. Abbondio. Lo spazio e il tempo. Tradizione storica e recupero architettonico*: 327–341. Como: Ministero per i Beni Culturali e Ambientali.

Cuervo-Cazurra, A., Martin de Holan, P. & Sanz, L. 2014. Location advantage: Emergent and guided co-evolutions, *Journal of Business Research* 67: 508–515.

CHCfE Consortium. 2015. *Cultural Heritage Counts for Europe, full report*. Retrieved from: http://bit.ly/2jERIwx.

Dann, N. 2004. Owners' attitude to maintenance, *Context* 83: 14–16.

Della Torre, S. 1999. "Manutenzione o "Conservazione"? La sfida del passaggio dall'equilibrio al divenire. In G. Biscontin & G. Driussi (eds), *Ripensare alla manutenzione*: 71–80. Venice: Arcadia Ricerche.

Della Torre, S. 2010. Preventiva, Integrata, Programmata: le Logiche Coevolutive della Conservazione. In: G. Biscontin & G. Driussi (eds), *Pensare la Prevenzione: Manufatti, Usi, Ambienti*: 67–76. Venice: Arcadia Ricerche.

Della Torre, S. 2018. The management process for built cultural heritage: Preventive systems and decision making. In K. Van Balen & A. Vandesande (eds), *Innovative Built Heritage Models*: 13–20. Leiden: CRC Press/Balkema.

Della Torre, S. 2019. A coevolutionary approach to the reuse of built cultural heritage. In G. Biscontin & G. Driusssi (eds), *Il Patrimonio Culturale in mutamento. Le sfide dell'uso*: 25–34. Venice: Arcadia Ricerche.

Della Torre, S. & Pili, A. 2019. Built Heritage Information Modelling/Management. Research Perspectives. In S. Della Torre, B. Daniotti & M. Gianinetto (eds), *Digital Transformation of the Design, Construction and Management Processes of the Built Environment*: 231–241. Cham: Springer International Publishing.

Fielden, B. 2003. *Conservation of Historic Buildings*. Third Edition. Oxford/Burlington: Architectural Press.

Haggis, T. 2004. Constructions of learning in higher education: metaphor, epistemology, and complexity. In J. Satterthwaite, E. Atkinson & W. Martin (eds), *The Disciplining of Education: New Languages of Power and Resistance*: 181–197. Stoke on Trent: Trentham Books.

Indirli, M., Kouris, L.A.S., Formisano, A., Borg, R.P. & Mazzolani, F.M. (2013). Seismic Damage Assessment of Unreinforced Masonry Structures After the Abruzzo 2009 Earthquake: The Case Study of the Historical

Centers of L'Aquila and Castelvecchio Subequo. *International Journal of Architectural Heritage: Conservation, Analysis, and Restoration* 7/5: 536–578.

Kallis. G. & Norgaard. R.B. 2010. Coevolutionary ecological economics, *Ecological Economics* 69: 690–699.

Lagerkvist, C. 2006. Empowerment and anger: learning how to share ownership of the museum. *Museum and Society* 4 (2): 52–68.

Laubichler, M.D. & Renn, J. 2015. Extended evolution: A conceptual framework for integrating regulatory networks and niche construction, *Journal of Experimental Zoology (Part B, Molecular and Developmental Evolution)* 324 (7): 565–577.

Magnaghi, A. 2017. La storia del territorio nell'approccio territorialista all'urbanistica e alla pianificazione, *Scienze del territorio* 5: 32–41.

Martinez, M., Sauleda, N. & Huber, G. 2001. Metaphors as blueprints of thinking about teaching and learning. *Teaching and Teacher Education*, 17: 965–977.

Mesoudi, A., Whiten, A. & Laland, K.N. 2006. Towards a unified science of cultural evolution, *Behavioral and Brain Sciences*, 29: 329–383.

Moioli, R. & Baldioli, A. 2018. *Dieci Anni di Conservazione Programmata*. Quaderni dell'Osservatorio di Fondazione Cariplo, 29.

Montella, M. 2003. *Musei e beni culturali: verso un modello di governance*. Milano: Electa.

Natale, M. 1998. Maestri e botteghe a Como nel Rinascimento: tre frammenti. In M.L. Casati & D. Pescarmona (eds), *Le arti nella diocesi di Como durante i vescovi Trivulzio*: 57–72. Como: Comune di Como.

Niles, D. & Roth, R. 2016. Conservation of Traditional Agriculture as Living Knowledge Systems, not Cultural Relics, *Journal of Resources and Ecology* 7(3): 231–236.

Norgaard, R.B. 1994. *Development Betrayed: the End of Progress and a Coevolutionary Revisioning of the Future*. London: Routledge.

Pereira Roders, A. & van Oers, R. 2011. Bridging cultural heritage and sustainable development, *Journal of Cultural Heritage Management and Sustainable Development*, 1/1: 5–14.

Petzet, M. 2010. *International Principles of Preservation*. Berlin: Bäßler.

Plevoets, B. & Van Cleempoel, K. 2013. Adaptive reuse as an emerging discipline: an historic survey. In G. Cairns (ed.), *Reinventing architecture and interiors: a sociopolitical view on building adaptation*: 13–32. London: Libri Publishers.

Plevoets, B. & Prina, D.N. 2017. Introduction. In Fiorani, D., Kealy, L. & Musso, S.F. (eds.), *Conservation-Adaptation*. EAAE Transactions on Architectural Education n. 65: 1–8. Hasselt: EAAE.

Pracchi, V., Capella, G.L., Capetti, A. & La Vecchia, C. 2010. La prevenzione in medicina e la prevenzione nel restauro. Stato dell'arte del dibattito: spunti e criticità. In G. Biscontin & G. Driussi (eds), *Pensare la Prevenzione: Manufatti, Usi, Ambienti*: 101–110. Venice: Arcadia Ricerche.

Prahalad, C. K. & Ramaswamy, V. 2004. Co-creation experiences: The next practice in value creation. *Journal of Interactive Marketing* 18(3): 5–14.

Prebensen, N. K. & Xie, J. 2017. Efficacy of co-creation and mastering on perceived value and satisfaction in tourists' consumption. *Tourism Management* 60: 166–176.

Quattrini, C. & Natale, M. 2017. Il Maestro dei Santi Cosma e Damiano. *Rivista Archeologica dell'Antica Provincia e Diocesi di Como* 199: 33–82.

Rescia, A.J., Perez-Corona, M.E, Arribas-Ureña, P. & Dover, J.W. 2012. Cultural landscapes as complex adaptive systems: the cases of northern Spain and northern Argentina. In T. Plieninger & C. Bieling (eds.), *Resilience and the Cultural Landscape. Understanding and Managing Change in Human-Shaped Environments*: 126-145. Cambridge University Press.

Ricoeur, P. 2005. *The Course of Recognition*. Harvard University Press.

Rypkema, D. 2012. Heritage Conservation and Property Values. In G. Licciardi & R. Amirtahmasebi (eds.), *The economics of uniqueness. Investing in Historic City Cores and Cultural Heritage Assets for Sustainable Development*: 107–142. Washington, DC: The World Bank.

Silberman, N. 2016. Changing Visions of Heritage Value. What Role Should the Experts Play? *Ethnologies*, special issue on Intangible Cultural Heritage, edited by L. Turgeon: 433–445.

Staniforth, S. (ed) 2013. *Historical Perspectives on Preventive Conservation*. Los Angeles: Getty Conservation Institute.

Therond, D. & Trigona, A. (eds) 2009. *Heritage and Beyond*. Strasbourg: Council of Europe Publishing.

Van Balen, K. 2015. Preventive Conservation of Historic Buildings. *International Journal for Restoration of Buildings and Monuments*, 21 (2-3): 99–104.

Van Balen, K. 2017. Challenges that Preventive Conservation poses to the Cultural Heritage documentation field. *The International Archives of the Photogrammetry, Remote Sensing and Spatial Information Sciences, Volume XLII-2/W5, 2017 26th International CIPA Symposium 2017, 28 August–01 September 2017, Ottawa, Canada*: 713–717.

Van Balen K. & Vandesande A. (eds) 2013. *Reflections on Preventive Conservation, Maintenance and Monitoring of Monuments and Sites*. Leuven: Acco.

Van Balen K. & Vandesande A. (eds) 2015. *Community involvement in heritage*. Antwerp-Appeldoorn: Garant.

Van Balen K. & Vandesande A. (eds) 2016. *Heritage counts*. Antwerp-Appeldoorn: Garant.

Van Balen, K. & Vandesande, A. (eds) 2018. *Innovative Built Heritage Models*. Leiden: CRC Press/Balkema.

Van Balen, K. & Verstrynge, E. (eds) 2016. *Structural Analysis of Historical Constructions – Anamnesis, diagnosis, therapy, controls*. London: Taylor & Francis Group.

Van den Bergh, J.C.J.M. & Stagl, S. 2003. Coevolution of economic behaviour and institutions: towards a theory of institutional change, *Journal of Evolutionary Economics* 13: 289–317.

Vandesande, A. 2017. *Preventive Conservation Strategy for Built Heritage Aimed at Sustainable Management and Local Development*. PhD. Dissertation, KU Leuven.

Vandesande, A., Van Balen, K., Della Torre, S. & Cardoso, F. 2018. Preventive and planned conservation as a new management approach for built heritage: from a physical health check to empowering communities and activating (lost) traditions for local sustainable development. *Journal of Cultural Heritage Management and Sustainable Development* 8, 2: 78–81.

Preventive Conservation - From Climate and Damage Monitoring to a Systemic and Integrated Approach – Vandesande, Verstrynge & Van Balen (eds)
© 2020 Taylor & Francis Group, London, ISBN 978-0-367-43548-6

Innovation and diversification of brick, Susudel – Ecuador

G. Barsallo, T. Rodas, V. Caldas & F. Cardoso
World Heritage City Project, School of Architecture and Urbanism, University of Cuenca, Ecuador

C. Peñaherrera & P. Tenesaca
Faculty of Chemical Sciences, University of Cuenca, Ecuador

ABSTRACT: The Later-Eris project, is developed within the framework of the World Heritage City Project of the University of Cuenca. This research focuses on the Susudel parish, located in the South of Ecuador, where it is observed that the majority of artisan producers offer them a single type of brick, known locally as a panelón (standard measurements 28x14x7 cm). This fact triggers a high competitiveness, accentuates the mono-production which affects an over-supply of an almost unique product and marketing at unsustainable prices, as this phenomenon unravels processes of unfair competition between producers. The objective of research is to innovate and diversify the production, from the floor covering brick, taking advantage of the potential and resources of the place, working alternatives with identity, based on the riches of the region. This is how, from the identification of the problem, a methodology of approach with artisans is proposed, which, more than the understanding of the productive dynamics and use of the cladding brick, promotes the conservation of the artisan production capacities, true tangible and intangible heritage in the study area. Additionally, the result is the innovation of designs and shapes, the characterization of the raw material and a joint work with artisans. Finally, a manual is generated that serves as a tool for the producers in order to be used as a reference.

1 INTRODUCTION

Susudel is located in the southern Andean region of Ecuador, where there is a clear vocation of artisanal brick production among its inhabitants, with 23% of the population that subsists by this activity (PDOT 2014). The territory has an excellent raw material, constituted by high quality lands and clays. However, despite the potential advantages offered by artisan brick for construction or other materials, its medium and long-term production capacity faces great challenges.

currently, most artisan producers offer them a single type of brick - called a panelón (28x14x7 cm.) - which is used regularly for the construction of walls and walls in urban contexts. This mono-production, triggers the over-supply of the same product, generating high competitiveness among neighboring producers and consequently the decrease of prices in the market. If this situation is maintained, the production and economic support of more than 100 families of the parish, linked to the production of artisan brick, is put at risk. In this context, the World Heritage City Project of the Faculty of Architecture and Urbanism of the University of Cuenca, in collaboration with the Faculties of Arts and Chemical Sciences, has articulated a research project, which seeks to develop alternatives that directly impact -in the improvement of the production of the tile used for the floor covering, diversifying it, innovating it and taking advantage of its potential and the resources of the place, working alternatives with identity, based on the cultural wealth of the region and focusing on the major regional markets such as Cuenca and Loja, fundamentally. The Project integrates several disciplines, including: architecture, chemistry, heritage conservation, history and arts, and close work with communities and local governments.

This is achieved through mixed methods with quantitative approach through literature review, surveys, in addition, research of samples of soil, pastes and bricks prepared by troughs, performing chemical, physical and mechanical analyzes.

Additionally, the research, projected from the qualitative point of view, applies an approach oriented from the socio-critical paradigm, in which the gathering of information takes place in the field, applying interviews and carrying out a continuous process of interaction with the community, with strong participation that allows the validation, return and diffusion of the information generated by the project, thus provoking a very active process of exchange of knowledge between artisans and technicians.

2 INFORMATION COLLECTION PROCESS

This process required a continuous approach towards the brickwork artisans, to understand the productive and marketing dynamics through participatory activities with recurrent field actions.

The objective was to determine, on the one hand, the degree or level of use of this material in the architecture of the parish and, on the other, to identify the forms and formats used in the different architectural components.

In the local context, some types of brick were registered and identified in the buildings surrounding the parish center of Susudel (New Susudel) in order to identify the level of brick use in the parish itself. Approximately 60 buildings were visited, detecting the use of this material in 26 real estate properties (43%). 31 records of elements were obtained, of which 23 of them corresponded to the use of brick in floors, bridges and steps.

In the international context, specialized literature and experiences related to the diversity of ways in which brick is produced were investigated. The results focused on identifying, above all, the types, dispositions, shapes and textures of the brick, as a basis for the formulation of the new proposals. Quoting some case studies like the Queche Laj product line. Developed in the "Valle Encantado" brick factory in el Tejar, municipality of the department of Chimalte-nago in Guatemala. Project that arises from the academy, in the Faculty of Architecture and Design of the Landívar University of Guatemala, with the purpose of giving alternatives to artisans so that they can produce innovative products both in their form and in their function (Pérez 2015). Other works are also analyzed, such as the Spanish Pavilion in Expo Zaragoza, whose work is part of an enterprise that, in its line of action, also includes the restoration of historic architectural elements, the use of color, organic forms and the integration of ceramics as a structural element of the building, thus achieving tradition, innovation, artisanal production (Mella 2012).

Another process that addressed the research was the characterization of Susudel's brick productive chain. This was done through the application of a survey applied in 56% of brickworkers belonging to different sectors of the parish.

The gathering of information considered components such as: supply, production, commercialization and consumption of the brick. At the same time, with visits to the brickworks, the database of types and formats of floor covering brick, which had been produced in pre-research initiatives, could be fed. It should be noted that the similarity of processes, the experience, and the knowledge inherited on the elaboration of the brick, allows the artisans to elaborate brick floor covering without major complications; however, the low demand for it makes continuous production impossible.

The field work allowed to know closely the productive process of the brick and to exchange knowledge with brick artisans. Once the productive chain was characterized, the process of designing the prototypes began. Some participatory workshops were developed in order to return the information received in the characterization phase of the productive chain and also worked together on the guidelines for new proposals; with these results in hand, the prototypes designed were disseminated and validated and finally the necessary adjustments to the proposed designs were made.

This exchange of experiences made it possible to discern criteria and reflections on the prototypes developed, presented and presented by the project's integrants, thus achieving the enrichment and improvement of the design process with the contributions of all those involved.

3 DESIGN PROCESS

After the rapprochement with the artisans and based on the study of the productive chain of brick for floor coverings, the high knowledge of ancestral empirical techniques with which artisans dominate production was confirmed. However, in spite of the limitations with respect to the new forms of designs, producers were identified that, at some point, had executed different models of bricks for coatings of square geometry. hexagonal and octagonal (Deleg 2010).

For the development of the design processes, in the first instance, participatory workshops involving teachers and students of the first cycle of the basic design subject of the Architecture career of the University of Cuenca were activated.

Young people contributed with basic concepts of two-dimensional organization for the generation of different shapes and patterns, from initial modules. However, these concepts were not compatible with the reality of brick for covering floors, since there were problems of modulation and production that made them unviable. That is why internal workshops are later held with the members of the Later Eris team. Based on the inputs generated by some students and focused on three principles of brick flooring innovation (texture, form and versatility), they worked and defined new concepts, which led to some initial design proposals.

Finally, in another process of self-interaction between the research team and the artisans, the prototypes of the molds (wood and iron) and pieces of wood were constructed, tested and validated.

3.1 Molds

One of the aspects discussed before the formulation of the new proposals was the design of the molds, mainly around the material to be used. At present, within the production processes of artisanal brick in Susudel, the use of traditional wood molds is standardized.

In contrast to the skepticism of artisans and the lack of knowledge of researchers on the use of metal molds, iron was used as the main option due to its durability, cost, physical properties, resistance and malleability, mainly. The selection of this material concentrated the criteria and recommendations of the multidisciplinary group of researchers from the University of Cuenca, technical and art areas.

The decision to opt for iron was a very positive factor for the innovation component, since the nature of this material allowed to expand the possibilities of designs and to make a leap from the orthogonal designs provided by the traditional wooden molds, to designs with organic shapes (curves). Tests were also carried out with molds made of plywood, with very poor results.

3.2 *Analysis of raw materials*

The contributions of the joint work with the Faculty of Chemical Sciences, were supported by a specific research through a thesis (Tenesaca 2019), which enriched the research from different analysis.

With the main objective of improving the dosing of pastes, to generate new brick designs for floor covering, based on the characterization of raw materials and pastes already used in the production of panelón bricks (Table 1- 2), 11 types of clays and 4 different pastes were studied, whose samples were taken from 4 brickyards located in 4 sectors of Susudel: San Gerónimo, Susudel Centro, Pullcanga and Sanglia. areas that were chosen and studied in the process of collecting information, with the advice of the Susudel community, bearing in mind the geographical distribution and the availability of the tundish to work together. Moisture, granulometry, plasticity, contraction, water absorption and pH tests were carried out.

It is necessary to determine the humidity because the tests carried out later require working according to the net weight of the material. In addition, the moisture content of the material gives an idea of the percentage of colloidal clay.

According to the method used, the particle size indicates the size of the particles of which the raw materials are composed. Generally, the larger the particle size, the smaller the amount of clayey material and the higher the content of minerals such as quartz, feldspar, carbonate, among others.

The plasticity indicates the ease or difficulty with which the piece can be shaped.

The greater the plasticity, the greater the formability. But this is opposed to an excessive contraction that can lead to cracks and breaks, so it is also necessary to

Table 1. Characterization of pastes from Susudel communities.

Community	Humidity	Plasticity	Total Contraction (%)	Percentage of fine aggregates (%)	Water absorption (%)	p^H
Pullcanga	24.70%	2.99	3.50%	40.02%	16.00%	5.74
St. Geronimo	20.00%	5.03	4.73%	44.81%	14.27%	5.95
New Susudel	21.40%	2.41	3.60%	31.08%	12.71%	6.93
Sanglia	28.50%	10.26	5.58%	50.08%	17.25%	6.68

Table 2. Characterization of clays from Susudel communities.

Community	Clay type	Humidity	Plasticity	Total Contraction (%)	Percentage of fine aggregates %	Water absorption (%)	p^H
Pullcanga	Black	5.46%	5.32	5.26%	51.64%	13.90%	6.15
	Red	6.77%	13.72	13.18%	59.59%	23.67%	4.83
	Ballast	9.11%	2.23	3.24%	19.91%	13.80%	6.54
St. Geronimo	Black	8.96%	7.25	8.88%	53.85%	16.97%	5.95
	Ballast	7.53%	3.32	3.48%	30.57%	14.15%	6.55
New Susudel	Plastic	4.20%	3.99	3.96%	36.19%	13.13%	6.55
	Sandy	5.23%	1.6	3.68%	28.28%	13.18%	6.73
Sanglia	Red	24.43%	10.83	9.98%	62.88%	20.47%	4.79
	White	37.84%	20.1	13.07%	79.73%	24.68%	5.53
	Ballast	10.68%	3.67	5.53%	35.20%	17.06%	5.68
	Sandy	16.54%	6.44	4.13%	43.08%	18.62%	4.75

determine this characteristic. Because plasticity must be balanced with shrinkage.

Water absorption tells me how porous the bricks are after burning. This physical process can be an indirect measure of their mechanical resistance, the higher the porosity, the lower the mechanical resistance.

On the other hand, it is necessary to determine the pH, since this can influence in some aspect the behavior of the clays during the process of elaboration of the products. Red clays have an acidic pH, and the pH value affects their mouldability. It is found that better results are obtained with pH values close to neutral (Singer & Singer 1979), so clays with very low pH values should not be used in higher proportions.

The adequate knowledge of the properties of the clays allowed to infer their behavior in the formulation of the pastes, and, therefore, in the pore for its constitution. While the characterization of the pastes served as a basis for comparison with the new ones that were obtained in the framework of this investigation. The results obtained are shown in the Table 1-2.

In order to determine if the pastes need to be optimized, a test was carried out using the pastes used to produce paneled brick. These pastes were tested in the production of brick for floors with the new designs obtained within the Later Eris project, using metal and plywood molds and following the same production process known to artisans.

This way, their behavior was evaluated in each stage, giving a rating of 4 for the best performance and 1 for the worst.

4 RESULTS

For its analysis, a database was created with the information gathered through the application of qualitative instruments, thanks to which the productive chain could be characterized in its entirety.

Finally, it should also be noted that this process seeks to promote a qualitative leap in the production of brick and in the systematization and good understanding of its production by developing a specific product such as the *"Manual de elaboración de nuevas propuestas de ladrillo"* for artisans of the sector, which reinforces its capabilities, and broadens its field of offers.

4.1 *Raw materials*

According to the analyzes carried out, it can be concluded that pastes from Pullcanga and Nuevo Susudel have a relatively low plasticity and shrinkage, while San Gerónimo has average values and that of Sanglia high, the latter presents the highest values in all aspects. The results show that the pastes taken from the Sanglia community brickyard does not need to be modified, while those from the communities of Nuevo Susudel, San Gerónimo and Pullcanga require a new dosage of its components, since the bricks showed signs of cracks and breaks (Table 3).

A Simplex-Lattice mix design was used for the new dosages (Gutiérrez & De la Vara 2012). For the selection of the pastes, the lowest percentage of water absorption and the highest modulus of flexion resistance are considered.

A mathematical model was developed for each of the proposed designs and these were analyzed by obtaining a response surface (Maldonado 2000).

The analysis of the characterization of the raw materials and of the pastes, have allowed to know the properties that they present; it is so that a suitable plasticity with values close to 6 avoids the appearance of cracks and crevices, and little resistance in green, a good distribution of the size of the particles favors the formation of the piece, having less porous products, and with a surface of better finish.

In addition, in order to complement the study of the pastes selected for the preparation of bricks for floor covering, a chemical analysis was carried out on the raw materials that make up each of the pastes that were modified, obtaining the results show in Table 4.

Table 3. Optimal mix determined for the bricks object of study of the communities of Susudel.

Community	Clay type			Water absorption (%)	Resistance to breaking (N)	Resistance module to flextion (N/m m^2)
Pullcanga paste	Black	Red	Ballast			
	16.66%	16.67%	66.67%	16.304	2638	3.189
OPTIMUM MIXTURE FOR						
New Sussudel paste	Plastic	Sandy				
	80%	20%		13.746	2232.66	2.444
OPTIMUM MIXTURE FOR						
St. Geronimo paste	Black	Ballast				
	20%	80%		13.764	2219.58	2.346

Table 4. Chemical analysis of the raw materials that make up each of the pastes that were modified.

Community				New Susudel				
Type of clay	SiO_2 (%)	Al_2O_3 (%)	Fe_2O_3 (%)	CaO (%)	MgO (%)	Na_2O (%)	K_2O (%)	lost to fire (%)
Plastic	64.41	17.68	2.71	0.77	0.39	3.61	4.41	6.03
Sandy	68.04	18.47	2.83	0.8	0.4	2.37	1.68	5.4
Comunity				St Geronimo				
Type of clay	SiO_2 (%)	Al_2O_3 (%)	Fe_2O_3 (%)	CaO (%)	MgO (%)	Na_2O (%)	K_2O (%)	lost to fire (%)
Black	71	15.74	2.85	0.43	0.29	0.65	0.85	8.19
Ballast	67.22	13.94	2.3	0.41	0.33	1.69	1.47	12.74
Comunity				Pullcanga				
Type of clay	SiO_2 (%)	Al_2O_3 (%)	Fe_2O_3 (%)	CaO (%)	MgO (%)	Na_2O (%)	K_2O (%)	lost to fire (%)
Black	65.51	17.43	3.06	0.74	0.85	2.12	1.18	9.10
Red	47.11	32.50	3.67	0.30	0.23	0.47	0.42	15.29
Ballast	72.07	13.74	2.71	0.41	0.25	0.88	0.96	8.99

Table 5. Rational chemical analysis of clays.

Community	Clays	Feldspars	Clays Substance	Free Silica	Fe2O3	Organic Material
Nuevo Susudel	Sandy	56.65	17.61	18.32	2.71	3.58
	Plastic	29.99	32.27	32.8	2.83	0.9
San Geronimo	Black	10.53	34.8	47.77	2.85	3.34
	Ballast	22.15	24.62	40.88	2.3	9.31
Pullcanga	Black	24.92	32.05	33.74	3.06	4.63
	Red	6.46	79.15	5.93	3.67	2.53
	Ballast	13.13	28.48	50.02	2.71	5.02

With the results of this chemical analysis we proceeded to perform a rational mineralogical analysis, which is a calculation that allows us to determine approximately the minerals present in each sample, as its name says it is a close calculation because it is not based on a mineralogical test (Table 5). With these analyses it was possible to observe that in the chemical composition of the clays of Nuevo Susudel, they are not so similar, although both behaved in a very similar way in the characterization, although the one is more plastic because it has a higher content of clay, it also presents considerable percentages of feldspars and free silica which explains in some way its behavior. With respect to San Geronimo, Black clay presents a greater quantity of free silica, whereas Ballast clay presents a greater content of feldspars, the clay substance content being the opposite, making Black clay more plastic. In Pullcanga clays, it can be seen that Black clay and Ballast clay are similar to the previous ones, that is to say, they are a mixture of clay, feldspar and free silica; on the other hand, Red clay presents a high content of argillaceous substance, which corroborates its greater plasticity.

At the end of the process, it was clearly possible to obtain pastes that facilitate the proposal of new brick designs for floor covering suitable for working with the new modular elements, tools and processes created in this project.

4.2 Prototypes and final design

The extensive information gathered in the field allowed us to verify that the artisans are in the capacity and the availability of new forms of bricks: As a result of the process, five new forms were obtained, inspired by the cos Andean vision, the material and inmaterial cultural wealth of the mountain region of Ecuador, with multiple possibilities of combinations, patterns and forms of placement.

The first design was the form called "Organic" form, arose from the sensation of wavy movement, against the conventional designs of orthogonal brick. This design produces sensations of continuity and diverse visual games, depending on the position of the observer. The design qualified as "Greek Cross" arose from the simplification and extraction of forms from the Chacana, which also represents a religious icon and a form adopted in the architectural plants of churches around the world. Other design was the form called "Rombo" was inspired by the textures of the wicker baskets, a material obtained from plant fibers of shrubs, used to weave baskets or different accessories such as carpets, furniture, etc. The design of the nominee "Chacana", is created by a religious icon of the Andean aboriginal peoples, which expresses the union or bridge between man and divinity.

Finally, the design of the nominee "Rosetón", was born from the representation of the jewelry of the women of the Ecuadorian highlands, particularly of the Saraguros ethnic group, the most important and the closest in the region (Figure 1).

In the present investigation it was possible to verify the excellent performance of the metal molds, unlike those of plywood, which, due to its consistency of industrial origin, deforms in contact with water, so that its behavior is not favorable in sustained production processes.

On the other hand, the use of metal in the production of the molds allowed the development of new innovation components suggested by the research team, such as the imprint and the organic form, which opens up promising possibilities whose limits are only in the Creative ability of the designers.

The elaboration of the metallic molds did not cause any type of inconvenience. The materials used for its elaboration, such as the platen and smooth circular rod, were fully adjusted to the needs of the proposed designs in both shapes and dimensions. The manufacture of the molds contemplated basic metal-mechanical works such as rolling (bending), drilling, cutting, welding and sanding. These works involved

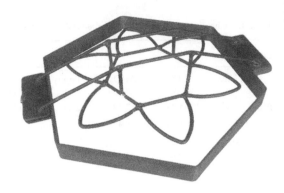

Figure 2. Mold proposal for handcrafted brick making.

docents and students from the Faculty of Arts of the University of Cuenca.

The advantages of metal molds over traditional wood molds are mainly the resistance, durability and momentum of the variety of shapes (linear or curved) that can be obtained (Figure 2). This undoubtedly has a positive long-term impact on the cost of this indispensable work tool for brick production and, above all, significantly expands design possibilities.

5 CONCLUSIONS

The development of new elements, of innovative forms and models, allows generating a series of simple and combined plots, using the raw materials themselves, adjusting to a procedure similar to the one traditionally used and matching the working conditions of the brick producers of the Susudel parish. The research developed has allowed the promotion of creativity and brick diversification through participatory processes, stimulating the recovery and transmission of the ancestral knowledge of the site, with a view to offering a new resource to improve the quality of life and social well-being of Susudel artisans.

The research has been supported by a multidisciplinary approach, with chemical, physical and mechanical analysis (scientific field of domineering rationality) as an important contribution to the design proposal (creative area of dominant intuition). These analyzes provided valuable data to increase the quality of the new design proposals, formulating adjustable brick types for the aesthetic and functional requirements demanded by traditional and modern architecture.

In addition, the formal search for the generation of molds and molds -which extend their useful life in relation to those commonly used, built with eucalyptus wood- has concluded that new materials such as iron, open new opportunities for the development of coatings of floors, opening to the artisans fields until the unexplored moment of innovation and creativity, which facilitates the diversification of their products.

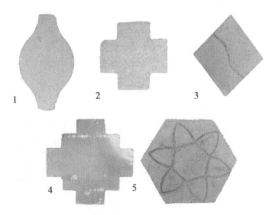

Figure 1. Brick proposal: 1) organic, 2) greek cross, 3) rombo, 4) chacana, 5) roseton.

With this research experience, the foundations for a future process of diversification of the designs offered by the brick industry in Susudel are laid.

The innovations in the various aspects developed for the generation of the prototypes obtained from the project are references for the production of new designs. These designs in turn can have an infinite diversity of sources of inspiration, in areas - to mention only a few - related to nature and the landscape, cultures and ancestral expressions, history and religions, abstract thought, geometry, the world of the ludic, etc. possibilities that challenge our own local capacities and that facilitate the materialization of the enormous potentialities that, thanks to creativity, can be developed.

REFERENCES

Deleg, N. (2010). Definition of a Semi-Industrial Bricks production process in the Susudel Parish. University of Cuenca.

Development Plan and Territorial Planning - Susudel (PDOT). Period 2009-2014.

Gutiérrez, H., & De la Vara, R. (2012). Analysis and design of experiments (Third Ed). Mexico City, Mexico: MC Graw Hill. ttps://doi.org/10.1017/CBO9781107415324.004.

Maldonado, O. (2000). Design of experiments with mixtures. University of Sonora. Retrieved from http://www.bidi.uson.mx/TesisIndice.aspx?tesis=8128.

Mella, J. M. (2012). The innovation in artesanal ceramics. (T. and C. Publications Center of the General Technical Secretariat, Ministry of Industry, Ed.) (Ed. I). Madrid.

Pérez, C. (2015). Free Press. Decae Sale In Ladrilleras de El Tejar. Chimaltenango.

Singer, F. & Singer, S. (1979). Enciclopedia de la Cerámica Industrial (Primera). Bilbao, España: Editorial Urmo.

Tenesaca, P. (2019). "Characterization of clays and pastes used in the artisanal brick making in the Susudel parish and its optimization for the manufacture of bricks for covering floors." University of Cuenca, Faculty of Chemical Sciences, Cuenca - Ecuador. http://dspace.ucuenca.edu.ec/handle/123456789/31735.

Preventive Conservation - From Climate and Damage Monitoring to a Systemic and Integrated Approach – Vandesande, Verstrynge & Van Balen (eds)
© 2020 Taylor & Francis Group, London, ISBN 978-0-367-43548-6

Monitoring of China's built heritage since 1950s: Historical overview and reassessment of preventive conservation

Wu Meiping
Key Laboratory of Urban and Architectural Heritage Conservation (KLUAHC) under Ministry of Education, Southeast University, Nanjing, China

Hu Shi & Li Xinjian
School of Architecture, Southeast University, Nanjing, China

ABSTRACT: This paper provides an historical overview of built heritage monitoring in China - based on four case studies since 1950s, including monitoring of two World Heritage Sites and two listed built heritage at the national level. Then it tries to assess how to effectively carry out the monitoring of built heritage in the context of preventive conservation, taking into account the technological development in built heritage monitoring, the evolution of World Heritage monitoring mechanisms, and the changes in conservation perception and strategies at China's national and local level.

1 INTRODUCTION

In ancient China, there was a preventive tradition: the attitude to periodically monitoring and maintaining its built heritage was preventive, as in "Nip it in the bud" in Chinese Medicine, and "prevent what seems to be the beginning of an unwholesome trend" in the education field.

Today, China's practice of monitoring its built heritage, while dating back to the 1950s and having been interrupted in the 1960-70s, since then 1980s has embraced advanced monitoring systems, as in unique projects such as the Yunyan Pagoda and the Maogao Caves.

Since 2000, China has gradually developed a national monitoring system for its World Cultural Heritage Sites. In 2006, in response to the monitoring mechanism of the 2005 World Heritage Convention, China's State Administration of Cultural Heritage (SACH) issued the MEASURES OF MONITORING AND INSPECTION OF WORLD HERITAGE SITES IN CHINA, while other World Cultural Heritage Sites began to carry out monitoring work in preparation for their periodic reporting to the World Heritage Committee.

In 2012, the Chinese Academy of Cultural Heritage (CACH) published the strategic document CONSTRUCTION PLAN FOR WORLD CULTURAL HERITAGE MONITORING AND EARLY WARNING SYSTEM 2013-2020, and initiated the Monitoring Center of World Cultural Heritage while pilot projects began in seven World Cultural Heritage Sites. In 2015, the Monitoring Center of World Cultural Heritage became officially established in CACH – to coordinate the monitoring work of China's World Cultural Heritage Sites.

The terminology of preventive conservation was introduced to the Chinese architectural heritage field in early 2000s. In 2006 a PhD research project on preventive conservation of architectural heritage was initiated at Southeast University (SEU). In 2011 an international conference on this subject was held at SEU in collaboration with the UNESCO Chair on (Preventive Conservation, Maintenance and Monitoring of Monuments and Sites, PRECOM³OS) of KULeuven. In 2014, the book PREVENTIVE CONSERVATION OF ARCHITECTURAL HERITAGE IN CHINA was published. Since then, the idea of preventive conservation has gradually attracted the official attentions at the national level and been written into the revised PRINCIPLES FOR THE CONSERVATION OF HERITAGE SITES IN CHINA, which states that "preventive conservation measures should be undertaken to reduce the need for interventions" (Article 12), and that "maintenance and monitoring are fundamental to the conservation of heritage sites" (Article 25). In 2017, the strategy - *"a shift from focusing on rescued conservation to paying equal attention to rescued conservation and preventive conservation"* has been defined in the 13th National Five-Year Plan for Cultural heritage Conservation (2016-2020) issued by SACH. Later this strategy was re-emphasized in the 2018 RECOMMENDATION ON ENHANCING THE CONSERVATION AND RE-USE OF CULTURAL HERITAG published by the State of Council, as one (of sixteen) important task to improve the Conservation Mechanism for Cultural Heritage, - *"supporting* the shift from focusing on rescued conservation to paying equal attention to rescued conservation and preventive conservation, from conservation of cultural heritage itself to the comprehensive conservation of both ontology and the surrounding environment". In late 2018, "Preventive Conservation", as a new expenditure type, was added to the national MEASURES FOR THE ADMINISTRATION OF SPECIAL FUNDS FOR CULTURAL HERITAGE CONSERVATION by SACH, which means a separate funds for preventive conservation from the national level.

2 AN HISTORIC PERSPECTIVE - FOUR CASE STUDIES

2.1 Case one: YUNYAN PAGODA monitoring

Yunyan Pagoda (Tiger Hill Pagoda or Huqiu Pagoda or 'Leaning Tower of China')(Figure 1), a national monument located in Suzhou, was completed in 961, with seven stories (47.7m) and an octagonal plan, built with a masonry structure designed to imitate the wood-structured pagodas prevalent at the time. It has gradually tilted, in more than one thousand years, due to forces of nature and its original foundation being half on rock and half on soil.

In 1950s, efforts were made to stabilize the pagoda and prevent further leaning. In 1955-56, an engineering-geologic survey was carried out, as well

Figure 1. Yunyan Pagoda, Suzhou, China.

as basic monitoring with regular observations of the tower's leaning and displacement to provide supportive data for the consolidation project in 1957. After an interruption in the 1960-70s, a precise measurement with photogrammetry was carried out in September 1979 by the Measurement Team in Jiangsu Provincial Architecture Design Institute, initially establishing a deformation measurement system to monitoring the tower's leaning and its displacement, foundation settlement, ground settlement and displacement around the tower. The measurement lasted for only one month but its technical way inspired the future systematic monitoring work.

October 1979 to January 1981: there was sporadic monitoring of cracks, tower leaning, displacement and ground settlement, carried out by a local team, providing data later considered useless due to its lack of specification requirements and the low quality of the monitoring equipment.

After January 1981: monitoring was done by an outside professional team - the Department of Measurement of Tongji University, whose improvements had the goals:

-of determining the deformation parameters under normal conditions (without construction disturbances) for the year preceding the consolidation projects and

–of collecting, tracking and analyzing the monitored data, to assist and control the safe construction of consolidation projects of the next two years: **December 1981 to August 1982** pile foundation engineering and **October 1982 to July 1983** the drilling grouting project.

Integral to their approach was monitoring of tools, provision of a fine adjustment device for the starting points and monthly monitoring of tower leaning, displacement and ground settlement. The pile foundation engineering proceeded smoothly during its first three months, yielding negligible changes to the tower leaning or deformation. The excavation pit was then speeded up from March to April 1982. During this period an obvious leaning change of 10 times normal was reported, as a warning, by the Tongji team, and the excavation pit was stopped. When it was restarted, by May 1982, the leaning speed had become much slower. In this way, the timely monitoring and data analysis ensured the safe construction and prevented the dangerous situations.

Since November 1982, due to the needs of high monitoring frequency, a local monitoring team was established under the restoration office of Yunyan Pagoda, to carry out alone, immediate monitoring to meet real-time needs of the on-going projects, and to cooperate with the Tongji team on the regular monitoring. Their monitoring collaboration continued until January 1984, both sides following the techniques established in 1979 but using more advanced tools, such as DSI Precision Level, T3 Theodolite, portable strain gauges and dial indicators, and establishing 53 observation points for ground settlement, 22 for cracks (vertical/oblique/horizontal/ground cracks) and 15 for deformation. The monitored data from both sides was mutually verified and a "Five-Fix-Criteria" (fixed person, fixed time, fixed tools, fixed station and fixed observation mark) was observed. Thereafter, a scientific monitoring system was installed. Since February 1984, the local monitoring team has been responsible for the entire monitoring including the pile foundation engineering during June 1984-May 1985 and the maintenance project of the first ground floor during May 1985-Sept 1985. In the subsequent monitoring during the test period October 1985- October 1986 the influence of the temperature changes to the ground uplift and tower expansion was gradually revealed. It was continued after the projects' completion with the same intensity and frequency for another 3.5 years (November 1986 – July 1990). From 1990-1995 monitoring was carried out by the same team but with a lower frequency of two to three times per year.

Since 1995, Suzhou Survey and Measurement Institute has been responsible for all the monitoring work, with a frequency of 2-3 time per year – allowing for special weather requirements related to major, potential structural damages such as the uneven settlement of the tower foundation and the tower displacement – as well as the regular monitoring for cracks in the consolidation areas within the tower.

Such monitoring has shown it to provide a reliable basis for the design, construction and quality analysis of consolidation projects.

2.2 Case two: Mogao Caves monitoring

Mogao Caves (Figure 2), a World Cultural Heritage Site located in Dunhuang, is one of the largest and finest Buddhist art centers, with 735 caves (the first cave excavated in AD 366), 46,000 m^2 of wall painting (5th – 14th century) and 2,415 stucco sculptures, spanning a period of more than 1000 years.

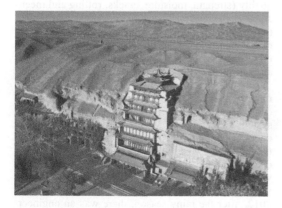

Figure 2. Mogao Caves (Main Entrance Tower), Gansu, China.

Monitoring began of the meteorological environment (temperature, relative humidity, rainfall, sunshine, wind speed, wind direction and sandstorms) in early 1960s - with a weather station and using simple instruments such as paper strip and calipers to monitor for changes of the wall painting and cracks in the rock mass – for basic information for the sandstorm control, wall painting conservation and consolidation of the rock mass..

In mid-1980s, four selected caves of different heights and sizes were monitored for interior temperature and humidity, but due to the low quality of monitoring equipment and the limitation of technical conditions, the monitoring intensity was very low and great contingent. Just as the monitoring in the 1960s, it was impossible to accurately and scientifically determine the characteristics of the sandstorm activities and changes of the other environmental factors.

Since the end of 1980s, a comprehensive monitoring system has been gradually developed covering environmental monitoring, micro-environment in caves, wall paintings, and cracks, and displacement of cave rock mass.

ENVIRONMENTAL MONITORING includes: a) 24-hour meteorological (temperature, relative humidity, rainfall, sunshine, wind speed and direction) by automatic weather stations; b) air quality (PM10, PM2.5) by continuously collecting samples daily (and once every six hours in sandstorm days), for elemental analysis with X-ray fluorescence spectrometry, for carbon analysis with a carbon analyzer and inorganic ion analysis with ion chromatograph; c) landform and surrounding environment by satellite imagery; d) sandstorm, to master the conditions for sand-lifting, the moving speed of the sand dunes upper surface, the relationship between the hardness of the top surface and the sand-lifting, the sand situation after establishment of the sand-preventing system, changes of the ground temperature after establishment of the forest belt, wind speed and sediment transport in different test; e) water environment, mainly on the changes in the flow and quality of the Daquan River, to determine whether the irrigation water in the front forest belt infiltrates into the cave area and whether the sand-fixing forest belt continuously penetrates to the sides and bottom, to understand the mechanism of soil moisture formation.

MICRO-ENVIRONMENTAL MONITORING inside the caves includes: a) for support of further conservation of the wall paintings: temperature, relative humidity, wall surface temperature, CO^2 concentration; b) for tourist management: tourists' influence on indoor temperature and humidity; c) water vapor distribution and migration in the rock mass. A smart monitoring system with wireless sensors was installed since 2006.

WALL PAINTINGS MONITORING includes: damage evolution (periodical inspection and monitoring in extreme weather), color changes and soluble salts in rock mass directly connected to the murals.

Although, originally, there were no inspection standards in place, neither description standards for recording damages nor color database, making it impossible to do any accurate assessment. Since then, there has been emphasis on the importance of research and training for damage types, characteristics and definitions, building up a specific color database for the wall paintings, exploring the hyper-spectral imaging technologies' application in the damage inspection and evaluation work. Cracks and displacement of cave rock mass are now monitored too.

The comprehensive monitoring being gradually built up today aims to help scientific environment control and implementation of other preventive actions for the caves, wall paintings and stucco sculptures.

2.3 Case three: Monitoring of Classical Gardens of Suzhou

Classical Gardens of Suzhou (Figure 3) are a group of nine gardens in the Suzhou region, spanning a period of 11th -19th century, with key features of classical Chinese garden design - constructed landscapes mimicking natural scenery of rocks, hills and rivers with strategically located pavilions and pagodas- which were added to the UNESCO World Heritage List in 1997 and 2000.

Monitoring commenced in 2004, aiming at preparing for the periodic reporting to the World Heritage Committee. In 2005, the Municipal Administration of Gardens set up an office with the name of "Monitoring and Management Centre for Classic Gardens of Suzhou". In 2006, the CONSTURCTION PROPOSAL FOR SUZHOU CLASSICAL GARDENS' MONITORING AND EARLY WARNING SYSTEM was drafted. In 2008 the proposed computer-based platform was implemented, with two levels of management (monitoring center and World Heritage sites), three database (historic archive, real-time information and early warning processing), and four levels of Information Links (World Heritage site - Provincial Administration of Cultural Heritage – SACH - World Heritage committee).

Included were periodic investigations for 11 categories, each with its own monitoring frequency:

Figure 3. Humble Administration's Garden, Suzhou, China.

buildings (semiannually), structures (semiannually), furnishings (once per season), vegetable (every two months), environment and buffer zone (semiannually), tourist flow (daily), safety measurement/regulations/technical ability of employees (once per season), historic documents (semiannually), and additional monitoring in unexpected conditions. In 2012 this became one of China's pilot projects for establishing World Cultural Heritage Monitoring and Early Warning System. For this, nine monitoring categories are included: 1 buildings, 2 structures and 3 furnishings, monitored by the Team of Ancient Garden Buildings under Suzhou Municipal Administration of Gardens; 5 environment, 6 buffer zone and 9 basic equipment by outside professionals; 4 plants, 7 passengers flow & 8 safety management, by the daily management offices of each garden. The associated 39 sub-categories of the above nine categories are individually monitored, as are the sub-items of ontology, natural environmental and human factors. (Refer to Table 1 & 2)

Monitoring of the Classical Gardens of Suzhou regarded as a comprehensive management monitoring system developed in the context of the monitoring mechanism of 2005 World Heritage Convention has, since then, been under improvement as an early warning system so as to support further preventive conservation activities. However, there is still a long way to go because, so far, there has been no scientific analysis of the monitored data.

2.4 Case four: Baoguo Temple Main Hall monitoring

The Main Hall of Baoguo Temple (Figure 4) built in 1013 is a renowned for its Dougong and timber frame structure. Monitoring of the wooden components and structure began in 2007 to collect data in order to understand better the damage mechanism involved and to eliminate potential risk over time.

The monitoring includes: both periodic and real-time monitoring of the structure (column leaning, settlement), regular monitoring of the wood material quality (strength, moisture, cracks, rotting and mechanical damages), 24-hour monitoring of environment (temperature, humidity, wind direction and speed, rainfall, surface and ground water, vibration), annual monitoring of ground settlement and displacement, and termite monitoring once or twice every season.

For indoor temperature and humidity, there were nine monitoring points, for structural settlement and displacement 32 points. This smart monitoring system with wireless sensors consists of 3 parts - data collection, management and display. In 2009, five automatic weather station were erected to collect data on outdoor temperature, humidity, rainfall and wind speed and direction. Work halted later for 5-6 years due to supervisor leadership replacement. In 2012, after the rainy season, there was an engineering-geologic survey and a structural health check. Work resumed at the end of 2015 and a proposal was

Table 1. Overall monitoring contents for Classical Gardens of Suzhou.

categories	sub-categories
1 buildings	1-1 buildings.
2 structures	2-1 garden rocks - artificial mountains; 2-2 garden rocks - single monoliths; 2-3 revetments; 2-4 flower stands; 2-5 bridges; 2-6 paving; 2-7 garden ornaments; 2-8 others.
3 furnishings	3-1 stools and chairs; 3-2 tables and desks; 3-3 couches and beds; 3-4 indoor hangings: hanging lanterns, inscribed boards, hanging panels; 3-5 indoor artworks: china, copperwares, standing screens, clocks; 3-6 calligraphies and paintings; 3-7 carved stones and stone inscriptions; 3-8 inscriptions on a tablet; 3-9 others.
4 plants	4-1 precious old trees; 4-2 landscape trees; 4-3 Penjing - miniature trees and rockery; 4-4 others.
5 environment	5-1 air; 5-2 weather; 5-3 water; 5-4 soil; 5-5 others: acid rain, dust, etc.
6 buffer zone	6-1 buildings within buffer zone; 6-2 sources of environmental pollution; 6-3 infrastructures; 6-4 others
7 passenger flow	7-1 passenger flow.
8 safety management	8-1 safety inspection; 8-2 accidents; 8-3 large exhibition and tourist activities.
9 basic equipment	9-1 electrical appliance; 9-2 fire equipment; 9-3 electrical equipment within the conservation area; 9-4 water supply and drainage system within the conservation area.

Table 2. Monitoring items and sub-items for buildings and structures of Suzhou Classical Gardens.

items	sub-items
Ontology monitoring	
B1. form and design	B1.1 appearance of built heritage; B1.2 attached heritage.
B2. material & structure	B2.1 wooden-frame structure; B2.2 stone columns, stone beams and lintels; B2.3 roof; B2.4 wall; B2.5 terrace; B2.6 garden rocks, bridges and other structures.
B3. location and environment	B3.1 position; B3.2 terrain environment; B3.3 landscape.
B4. traditional crafts	B4.1 traditional crafts
B5. spirit, connotation and intangible heritage	B5.1 spirit, connotation and intangible heritage
Monitoring of natural environmental factors	
E1 micro-environment	E1.1 temperature; E1.2 humidity; E1.3 wind; E1.4 rainfall; E1.5 sunshine; E1.6 illumination
E2 air	E2.1 dust & soot; E2.2 air pollutants.
E3 water and soil	E3.1 surface water; E3.2 underground water; E3.3 soil.
E4 biological attack	E4.1 termite; E4.2 other insects; E4.3 animal wastes.
E5 natural disaster	E5.1 surface subsidence; E5.2 meteorological disasters; E5.3 other natural disasters.
E6 tree and plant	E6.1 tree; E6.2 grass and other plant.
Monitoring of human factors	
H1. Maintenance	H1.1 maintenance measures against water and moisture damage; H1.2 maintenance of electrical appliance and equipment; H1.3 cleaning work.
H2 improvement of environment	H2.1 planting; H2.2 water related; H2.3 roads.
H3. repairing and restoration	H3.1 repairing and restoration
H4. tourist management	H4.1 tourist flow; H4.2 vandalism of tourists; H4.3 tourist safety.
H5. Employee management	H5.1 permanent employee management; H5.2 temporary employee management
H6. Security management	H6.1 fire-fighting; H6.2 alarm system; H6.3 lightning protection; H6.4 emergency responses
H7. Use and function	H7.1 function; H7.2 improper use.
H8 vandalism	H8.1 vandalism
H9 conservation projects	H9.1 damages caused by conservation projects
H10. management system	H10.1 management system

initiated for upgrading the original monitoring system. The monitoring points for indoor temperature and humidity were increased from 9 to 30. Upgrading of the monitoring equipment is on-going and new automatic monitoring is being carried out for the settlement and leaning of key components and for air pollutants, such as SO^2, CO^2, PM2.5, etc.

Monitoring of the Main Hall was implemented to test the idea of **using modern technology to do preventive monitoring** before the building experienced damages. However, because of the lack of a comprehensive evaluation of the main risk factors before the establishment of the monitoring system, most of the monitoring work was limited to superficial factors, such as the temperature, humidity, displacement, deformation, etc. As well, the monitoring data obtained was not analyzed nor was there researches on the relationship between different risk factors and damage changes.

In this case the entire monitoring system can be regarded as only a scientific prescription without scientific benefit, which is not consistent with the

Figure 4. Main Hall of Baoguo Temple, Zhejiang, China.

original intention. Such casual monitoring fails as a scientific method for understanding damage mechanism, and as an effective basis for any preventive conservation measure.

3 FOUR CASES COMPARED AND CURRENT CHALLENGES

3.1 *Four cases compared*

The first two cases are the early examples of supportive monitoring, for consolidation at Yunyan Pagoda from 1955, and for environmental control at Mogao Cave from early 1960s. Suzhou Classical Gardens' monitoring, initiated in 2004 in the context of the World Heritage Monitoring Mechanism, and developed in the background of China's efforts at establishing the World Cultural Heritage Monitoring and Early Warning System, has recently also influence the monitoring of the Mogao Caves. The fourth case Baoguo Temple is an attempt to use monitoring in order to better understand the damage mechanism and thus support the preventive conservation activities.

Those four cases illustrate the evolution of built heritage monitoring in China since 1950s, reflecting the technical development, the response to the World Heritage Monitoring Mechanism, the national strategy for the monitoring of WH and national cultural heritage, and the influence of the new idea of preventive conservation.

Based on the four case studies and current monitoring practices, we can see that the influential factors for monitoring of built heritage include: technical development, financial support, policy at the national and international level, and new conservation theory and ideas - preventive conservation etc.

3.2 *Problems and challenges facing the current monitoring practices in China*

Although there have been good outcomes for both single structures and complex heritage since the 1950s, and there is now in place a national strategy with strong financial support for the monitoring of WH and important cultural heritage, there are still many problems and challenges to be faced.

The lack of systematic fundamental research on external environmental risks, properties of the structure and material, damage types and mechanisms, makes it difficult to define the essential factors to be continually or periodically monitored. Such research usually needs a long-term plan and takes a lot of time, which is inconsistent with the existing performance evaluation system and the unstable terms of supervisory leadership.

The lack of analysis of the monitored data results in the built comprehensive and smart monitoring system with high technological equipment capable of monitoring everything, being only a scientific prescription but yielding no physical benefits. Hindering analysis work is the difficulty in setting out common methods since heritage sites are so various in their aims, methods and requirements, and also the fact that since it is often regarded as a type of research work, specific funding for it is hard to arrange.

Today it is not easy to set up a separate qualification control for the fast-growing monitoring field which since 2015 has grown into an independent industry with its own specific responsibilities. A monitoring project generally consists of two stages: 1) preparing for the monitoring proposal, which is carried out by companies meeting the specific qualification requirement for survey and measurement for cultural heritage, and 2) the on-site arrangement of the monitoring equipment by professional companies, for whom it is not mandatory to meet qualification requirements, unlike the case for other types of conservation activities.

The team responsible for drafting the proposal normally chooses the specific monitoring equipment, although, theoretically, it is only necessary for them to point out the parameters required for each items of monitoring equipment required. Most such equipment is made in China with some being assembled with imported sensor chips and probes, which are always very expensive. Interference by underground interests which could affect or even destroy the real aims of the monitoring projects requires that clear supervision and control be stipulated.

Qualified management and control of the monitoring is wanting, monitoring of the ordinary built heritage is minimal although it forms the majority of the country's built heritage, and there is no coordination between monitoring, maintenance and management, which plays a key role for the success of preventive conservation in the field of built heritage.

4 MONITORING DESCRIBED IN THE CONTEXT OF PREVENTIVE CONSERVATION

4.1 *Preventive conservation definition*

There is a generally accepted definition for preventive conservation in the movable heritage field, issued by ICOM-CC in <Terminology to Characterize the Conservation of Tangible Cultural Heritage> in 2008, in which it is explicitly stated that "Conservation embraces preventive conservation, remedial conservation and restoration". It helps us to understand the role of preventive conservation under the umbrella term "conservation", although the given definition of preventive conservation is not suitable for the built heritage field. Preventive conservation in the field of built heritage should be understood differently from that of either museum objects or archaeological sites, since it is impossible to change or optimize the environmental conditions while built heritage are exposed to the natural environment and often required of new adaptive functions.

Based on the limited literature held by the author, preventive conservation as a professional terminology appeared in the introduction of Garry Thomson's book THE MUSEUM ENVIRONMENT in 1978, and was introduced to the field of archaeological sites in 1990s (Mike Corfield 1996) and later, to the field of architectural heritage in early 2000s (Koenraad Van Balen 2006). There have been other terminologies, such as preventive restoration (Cesare Brandi 1963), programmed conservation (Giovanni Urbani 1976), prevention of deterioration and preventive maintenance (Bernard M. Feilden 1979, 1982, 1994, 2003), planned conservation (Stefano Della Torre 1999), planned preventive maintenance (Brian Wood 2005), preventive and planned maintenance (Paolo Gasparoli, Roberto Cecchi 2010), preventive and planned conservation (Stefano Della Torre 2013, 2018), etc.

In the past decade, there are two representative perspectives for preventive conservation of built heritage: 1) Preventive and Planned conservation (PPC), theorized by Stefano Della Torre since 1999, starting from Urbani's programmed conservation (conservazione programmata) in 1970s and developed in 2000s, is a holistic and long-term strategy with aims of preserving material authenticity, the management of transformations, the effectiveness of both conservation activity and expenditure for conservation, and the integration between conservation and valorization; 2) Preventive Conservation (PC), theorized by Koenraad van Balen since early 2000s, based on the researches on the damages diagnosis of structure and material since 1980s and the successful practices of Monumentenwacht Vlaanderen since early 1990s, is a holistic system integrating monitoring, maintenance and other interventions, to promote professional periodic inspection and proper maintenance, to raise public awareness of the importance of preventive conservation and encourage sustainable management. Those two, with long-term academic collaborations of mutual influence and cross-infiltration, have laid the foundation for today's research on preventive conservation of built heritage.

Currently there is still not a clear definition for preventive conservation of architectural heritage, what is widely accepted that it should be based on three levels of prevention, analogies to the prevention idea in the medical profession, i.e. firstly to avoid causes of unwanted effect to act (degradation of health), secondly to monitor for early detection of symptoms caused by unwanted effects, thirdly to avoid further spread of the unwanted effect or the generation of new unwanted (side) effects" (Stefano Della Torre 2010, Koenraad Van Balen 2015).

The definition offered by the author, based on the research and practices at the international level, combined with the architectural heritage conservation system in China, is as follows: preventive conservation of built heritage relates to all means and measures for delaying deterioration and avoiding the causes and further spread of damages - through identification, investigation, evaluation, tracking, preventing and controlling all the risk factors - advocating planned prevention and minimum intervention in order to best preserve the authenticity, integrity and continuity of built heritage. It is a holistic and long-term care system composing the actions of: risk assessment and management, specific disaster prevention planning at the regional level, damage diagnosis of structure and material, conservation design to reduce the vulnerability of heritage building against disasters, regular inspection, monitoring, maintenance and other planned intervention, and corresponding safeguard mechanisms, etc.

4.2 Monitoring in the context of preventive conservation

Monitoring is regarded as a main operational tool for preventive conservation. It helps early the detection of symptoms caused by unwanted effects (referring to the three levels of prevention in medical analogy), assists the previous risk assessment and scientific evaluation, tracks and ascertains the effectiveness of the methods used, and ensures the long-term performance of the structure and any other materials used.

Due to either external environmental risks or internal deterioration or decay of structure and material, monitoring can be specific for natural disaster (geological and meteorological disaster, atmospheric pollutants...), for deterioration factors and damages changes (settlement, displacement, termite erosion...), and for dangerous activities affecting the structural safety performance (underground mining, traffic vibration, safe construction ...).

In the framework of preventive conservation, monitoring should be carried out before, during and after intervention activities. What is more important thing is to establish a coherent strategy which combines monitoring, periodic inspection, maintenance and daily management, wherein monitoring work could be carried out real-time (for very limited items) or continually or regularly as periodic inspection, maintenance work could be preventive or curative, and daily management could be more effective by sharing the knowledge of monitoring, periodic inspection and proper maintenance among internal staffs/workers and external users. The aim of such coherent strategy would be to delay the needs for larger interventions, thus ensure the authenticity and integrity of built heritage - which is the essential idea of preventive conservation.

In Chinese context, the current monitoring of WH sites and important national built heritage, usually more readily financed and supported by government, tends to build a comprehensive monitoring system and promote real time monitoring with high technological equipment - easily produces huge amount of monitoring data. Such comprehensive system should be developed into an information management system and provides accessible dialogues with the management teams, for the planning of conservation activities in order to shift from urgent repair/restoration to planned conservation, for controlling costs and evaluating the effectiveness of both the conservation activities and costs in a long term. In addition, more attention should be paid to the previous fundamental research (on structures/materials properties, damage mechanism and external risks, etc.) and the afterwards analysis of the monitoring data, wherein the former helps to scientifically define the essential factors needed to be continually or periodically monitored, and the latter helps to better understand the evolution and decide the possible responding preventive activities.

For common built heritage, different from those for WH sites, any comprehensive monitoring system is neither suitable nor necessary, and it is essential to first define influential factors and then to carry out periodic monitoring and inspection at low frequency and cost, by professional organizations. Special provincial or local organizations could be founded separately, outside or within the existing conservation institutions, perhaps follow the model of Monument Watch (Monumentenwacht) in Netherlands and Belgium, which carries out the periodic inspection of built heritage in order to raise awareness among owners and caretakers of the importance of proper maintenance and preventive conservation.

5 CONCLUSION

When we look back at the monitoring practices of the past few decades in China, we see that monitoring, as a scientific measure, can assist and ensure the success of conservation projects. But, recent monitoring practices since 2000, especially for WH and other important national cultural heritage, in responding to the World Heritage Monitoring Mechanism and to China's own monitoring strategy, have focused on building up a comprehensive monitoring system and real time monitoring with high technological equipment. It may not be cost-effective nor suitable nor necessary for common built heritage.

From the perspective of preventive conservation, the existing comprehensive monitoring systems of WH could go further as an information system able to dialogue with the management system in the direction of planning, controlling and evaluating various conservation activities at different stages. Real time monitoring with high technological equipment should only be encouraged for individual cases when it is really necessary, and should not overshadow the effectiveness of periodic inspection at low frequency and cost, which may be more suitable for the common built heritage. It is more important to establish a coherent strategy combining monitoring, periodic inspection, maintenance, daily management and other planned conservation activities. Such coherent strategy may be suitable for all types of built heritage.

ACKNOWLEDGEMENTS

This work was carried out within the scope of the research project "Prevention conservation of immovable cultural heritage: the international theoretical evolution and successful practices" funded by SACH. The authors convey their sincere gratitude to Mr YAO Chen, LING Ming and YE Simao from SACH for their supporting the current research activities to Prof. SHEN Yang for his support and help.

REFERENCES

ICOMOS China. 2015. Principles for the Conservation of Heritage Sites in China. Beijing: Cultural Heritage Press.

ICOM-CC. 2008. Commentary on the ICOM-CC Resolution on Terminology for Conservation. http://www.icom-italia.org.

Boniotti, C., Konsta, A., Pili, A. 2018. Complex properties management: Preventive and planned conservation applied to the Royal Villa and Park in Monza. Journal of Cultural Heritage Management and Sustainable Development, Vol.8 No.2: 130–144.

Brandi, C. 2015. Nardini Editore (translator). Theory of Restoration. Rome: Piazza della Repubblica.

Corfield, M. 1996. Preventive conservation for archaeological sites. Studies in Conservation. Volume 41:32–37.

Della Torre, S. 2013. Planned Conservation and Local Development Processes: The Key Role of Intellectual Capital. In Koenraad van Balen & Aziliz Vandesande (Eds). Reflections on Preventive Conservation, Maintenance and Monitoring of Monuments and Sites. Leuven: ACCO.

Della Torre, S. 2018. The management process for built cultural heritage: preventive systems and decision making. In Koen Van Balen & Aziliz Vandesande (Eds). Innovative Built Heritage Models. Leiden: CRC Press.

Feilden, B. 1982. Conservation of Historic Buildings. London: Butterworth &Co (Publishers) Ltd.

Gasparoli, P., Cecchi, R. 2012. Preventive and planned maintenance of protected buildings. Methodological tools for the development of inspection activities and maintenance plans. Firenze: Alinea Editrice.

Gordon, R. S. 1983. An operational classification of disease prevention. Public Health Rep, 98(2):107–109.

Meiping, W. 2014. Preventive Conservation of Architectural Heritage in China. Nanjing: Southeast University Press.

Meiping, W. 2019. The Monument Watch model for preventive conservation of immovable cultural heritage: a case study of Monumentenwacht Vlaanderen in Belgium. Preventive Conservation Forum of Heritage Preservation International, Shanghai, 31 October – 02 November, 2019.

Meiping, W., Guangya, Z., Shi, H., Jianguo, W. 2013. Preventive Conservation of Architectural Heritage under Chinese Current Conservation Scheme. In Koenraad van Balen & Aziliz Vandesande (Eds). Reflections on Preventive Conservation, Maintenance and Monitoring of Monuments and Sites. Leuven: ACCO.

SACH. 2017. The 13th National Five-Year Plan for Cultural heritage Conservation and Development. http://www.sach.gov.cn.

Thomson, G. 1978. The Museum Environment. Boston: Butterworths.

Van Balen, K. 2015. Preventive Conservation of Historic Buildings. Restoration of Buildings and Monuments, 21(2-3):99–104.

Vandesande, A. 2017. Preventive Conservation Strategy for Built Heritage Aimed at Sustainable Management and Local Development. Thesis to Obtain the Degree of Doctor in Engineering. KU Leuven.

Wood, B. 2005. Towards innovative building maintenance, Structural Survey, Vol. 23 No. 4: 291–297.

Zong, T. 2015. Upgrading proposal for the monitoring system for Main Hall of Baoguo Temple. A presentation at the Ancient Buildings Museum at Baoguo Temple, Ningbo.

Preventive Conservation - From Climate and Damage Monitoring to a Systemic and Integrated Approach – Vandesande, Verstrynge & Van Balen (eds)
© 2020 Taylor & Francis Group, London, ISBN 978-0-367-43548-6

A brief review on preventive conservation and its application in China's conservation background

Qingwen Rong
School of Architecture, Southeast University, China

Baoshan Liu
Beijing WMWB Culture Technology co., Ltd, China

Jianwei Zhang
School of Archaeology and Museology, Peking University, China

ABSTRACT: Preventive conservation (PC) can be seen as the epitomes of the "scientific phase" in architectural conservation after the Venice Charter. Although the idea of PC in the field of architecture can be traced back to John Ruskin in the 19th century, later William Morris, and even implied in the earlier idea of conservation history, its main development is after middle 20th century. Through the review of PC's development process, this paper briefly reviews the progress of PC researches and practices in architectural heritage in an International level, to sort out the past and existing related organizations and their main activities. By understanding relevant scholars' research and practice, this paper argues that the meaning of PC is more than "prevention". PC ideas is applied to China's conservation work in the form of working package modules, in order to offer several reflections on PC's application in diverse policy, management and culture backgrounds.

1 A BRIEF REVIEW ON THE PROGRESS OF PC RESEARCHES AND PRACTICES IN ARCHITECTURAL HERITAGE AT AN INTERNATIONAL LEVEL

It is known that PC is associated with multiple terms (Table 1). To understand the beginning of the concept of "PC" in architecture, two important scholars must be mentioned here: Cesare Brandi, the leader of the conservation discipline of the 20th century, and Giovanni Urbani (one of the former Deans of Italian Istituto Centrale del Restauro). In the 1950s, Cesare Brandi introduced the term "Restauro Preventivo" into the field of architecture (which could be regarded as the first most important introduction of "Preventive/Preventative" in the field of architecture); "Conservazione Programmata" was proposed by Giovanni Urbani in the 1970s (Della Torre 2013), which emphasizes the programmed innovation on conservation approaches (Della Torre 2010a), the conservation horizons regarding heritage as a whole with its territory and integrated thinking. Monumentenwacht (MOWA) in the Netherlands (from 1973) and "Conservazione Programmata" in Umbria in Italy (from 1975-1976) represented the two main original lines at the beginning of development of PC practices in 1970s (Figure 1).

In the decades of PC's development, its theory and practice have yielded fruitful results. Particularly in the past decade, a large number of international research institutions and practical projects for preventive conservation of architectural heritage have constantly emerged. With the growth of theories in this field, many new methodologies and tools have also been developed.

Table 1. The terms used in PC researches in built heritage.

Preventive/ Preventative	Preventive Conservation (*Conservazione Preventiva* in Itaian) Preventative Conservation Preventive Restoration (*Restauro Preventivo* in Italian) Preventive Preservation Preventive Maintenance
Planned/ Programmed	Planned Conservation Programmed Conservation (*Conservazione Programmata* in Italian) Planned Preventive Conservation, Planned Preventive Maintenance, Preventive and Planned Conservation, Preventive Planned Conservation
Other related terms	Predictive Conservation/Maintenance, Proac-tive Conservation/Maintenance, Integrated Conservation

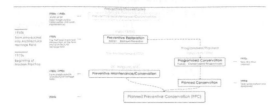

Figure 1. The two main original lines at the beginning of development of PC practices.

Figure 3. The main international organizations involved the practice of PC.

1.1 The main international organization involved in the practices and researches of PC

The current international organizations involved the practice of PC mainly include Monumentenwacht and some countries' conservation organizations who have learned from its model, the PRECOM3OS Chair of the University of Leuven and the extensive cooperation network around it, Department of Architecture, Built Environment and Construction Engineering, Politecnico di Milano, Italy and its many collaborators, UNESCO Chair of "Cultural Heritage and Risk Management" at Ritsumeikan University, Japan and its partners Kyoto University, Waseda University (they have carried out a series of work on risk prevention), etc. PC research was also carried out by institutions from the University of Minho and Universidade do Porto in Portugal, the University of Salamanca in Spain, the University of Limoges in France, the University of Carleton in Canada, and countries such as Brazil, Chile, and China. The distribution of the main institutions conducting PC research is shown in Figure 2.

Figure 2. The related research projects of PC in built heritage or the risk prevention, monitoring of PC in built heritage on the series of frameworks in the EU-funded R&D Framework Programme FP1-8.

1.2 The research projects on PC

The PC-related series of frameworks in the EU-funded R&D Framework Programme (FP) includes the Wood-Assess project under the EU's Fourth R&D Framework Programme (FP4) (1990s), and related studies in the FP6, FP7 and so on (Figure 3).

In addition to the EU series of R&D frameworks, there are many remarkable major international cooperation projects in recent years including VLIR-CPM, CHANGES, HeritageCARE, Art-Risk, CPRE and so on.

1.3 The development of PC tools

With the extension of the theory, a number of new tools and methods have also been developed, especially in the last decade (Table 2). Special attention is paid to the development of tools such as model algorithms and software. It is closely related to the main work content of the PC - emphasizing the application on scientific and technological measures in daily monitoring, data accumulation, diagnosis and prediction, and disaster prevention and mitigation.

1.4 Conferences and related work

Judging by the increasing quantity of PC-related research literature, after 2000, the development of PC theory and method has synchronized with the new round of conservation paradigm in a new stage. Relevant international conferences have been constantly organized and the number of PC related articles has increased significantly (Figure 4). From the emphasis on routine maintenance and regular

Table 2. The related research projects of PC in built heritage or the risk prevention, monitoring of PC in built heritage on the series of frameworks in the EU-funded R&D Framework Programme FP1-8.

Names of tools	Reference
S.I.R.Co.P.	(Benatti et al. 2014)
Planet Beni Architettonici	
MDDS	(Core 2009)
MDDS-COMPASS	
MDCS	(De Rooij 2016)
MAKS	(Stulens et al. 2012)
COMEET	
FBSL	(Ibáñez et al. 2016)
MPlan	(Ferreira 2018)
CityGML	(Zalamea Patiño et al. 2016)
CityGML-ADE	
MHS	(Chiriac et al. 2013)
WETCORR	(Henriksen et al. 1998)
GISWOOD	(Haagenrud et al. 1998)

Figure 4. Statistics on the built heritage PC related publications in English (counting up to 2018.03, some case-study articles are not included here).

* Only the books and articles particular in the built heritage PC study are counted, and omitted the book review. There are also a large number of documents discussing PC-related maintenance, monitoring, and disaster prevention work, as well as specific research on buildings' interior PC, which are not counted. For some publications that only have a few chapters about PC-related content are not counted in this paper.

inspections in the 1970s to the inclusion of attention on disaster management and risk prevention in the 1990s, and to recent years, in parallel with the whole progress of heritage conservation research, PC-related research has gradually introduced many top issues under the new round transformation of the conservation paradigm such as community and cost efficiency (Professor Koenraad van Balen and Dr. Aziliz Vandesande have many publications on these topics) (Della Torre 2010b, Vandesande 2017 & Vandesande et al. 2018). At the same time, some scholars' interests focus on the PC's programme package, organization management and legal support.

2 THE CONCEPTUAL CONNOTATION OF PC IN BUILT HERITAGE FIELD

After reviewing the progress of International PC researches in built heritage field, we tried to further understand the conceptual connotation of PC. PC has been defined by several famous scholars (e.g. Sir Bernard Melchior Feilden) and councils (e.g. "Conservation embraces preventive conservation, remedial conservation and restoration" by ICOM-CC) (ICOM-CC, 2008). Also, the practices have been developed effectively by organizations like MOWA with the "umbrella" structure. However, there are still misunderstandings in some discussions on the conceptual connotation of PC. Therefore the connotation of PC is clarified in three phases here basing on some valuable references: first, the difference of same-name concept between the field of architecture and other fields. It has already been clearly explained by Professor Koenraad van Balen. Second, the relationship between "PC" and "prevention", here can be referred to the discussions by Professor Salvador Muñoz Viñas et al. Third, the significance of environmental control for the PC in architectural heritage field. Cesare Brandi has given some keen insights of this point in his great work-*Teoria del Restauro*.

ICOM-CC has clearly defined the "indirect" characteristics of PC for movable heritage in 2008. Professor Salvador Muñoz Viñas's classification is also straightforward. According to his point of view, except the "informational preservation" that does not actually apply on the conservation target itself, 'preservation' can be divided into two categories: the first one is to apply on the conservation target itself, which is called 'direct preservation' that causes some invisible changes on the conservation target's attribute and making an effect in the limited time; and the other one "environmental preservation" that acts on the environment where the conservation target is located, changes the environmental attributes. Theoretically speaking, it could sustainably make contributions. It won't change the visible and invisible features of the object and can continue to work indefinitely (Muñoz Viñas, 2015). In terms of the movable heritage, PC is "environmental preservation". Obviously, the PC in the built heritage field does not match the above features. Therefore, the definition of its concept is different. Professor Koenraad van Balen has clarified the definition of PC in the field of built heritage, archaeology and the field of museology (2015).

Secondly, in the field of movable cultural heritage, Professor Salvador Muñoz Viñas pointed out that the definition of "preventive preservation" often fails to clarify the key difference between PC and conservation methods in general, as most of conservation methods aimed at preventing the future deteriorations (2005). Similarly, the definition of PC in architecture often fails to be distinguished from the definition of "conservation" in general which is confusing (For example: defining PC as any conservative action for preventing potential damage on the value of built heritage, material degradation or structural degradation and etc.) (Muñoz Viñas, 2005). In fact, Professor Stefano Della Torre has accurately revealed its essential characteristics in the discussion of Planned Conservation: "its deep character has been found in the long term vision applied in any action on cultural heritage, including planning of large scale Features" (2013, 123-127). Compared with the emphasis on "prevention", the key connotation of the concept PC in built heritage is its ideological structure that is distinguished from the general conservation method. PC is not only a concept, but rather a way of thinking and a comprehensive framework. To judge whether a conservation project is "PC" or not, the key point is not whether it is committed to the possible damage in the future, nor whether the actual situation of the object it faces, nor whether it adopts the judgment on the mechanism of material degradation or regular monitoring and other measure. Compared with the passive and lagging response to certain problems, the key point is initiative and pre-planning an inclusive, comprehensive, cyclical and long-term working framework.

Third, although as mentioned above, it is difficult for PC in built heritage field to conduct the environmental control like museum collections and archaeological sites, this does not mean that environmental control is not important in the PC of built heritage field. In addition to the possible ways for reducing environmental risks such as reduce pollution, clean up standing water, and manage (remove or introduce) vegetation, control traffic, noise and vibration, etc.; the design intent of a historic building as an aesthetic work and a historical record, and the historical connection between a historic building and its context and "*intangibilità* (intactness, integrity)" of its surrounding areas need to be particularly considered as well. This insight has been pointed out by Brandi in the introduction of the concept of "Restauro Preventivo"(2005). In his great work ***Teoria del Restauro***, Brandi uses the example of Sant'Andrea della Valle to illustrate the negative impact of changes in the historical structure of the urban environment in which the building is located on its artistic effect, and argues that the preventive restoration should not be restricted to the consideration of architecture itself, but also to the requirements of surrounding area (2005). From Brandi's idea, we should realize that the treatment of the built environment is an unavoidable question to define the field of PC research precisely in the contemporary era. In other words, in addition to the objective analysis in the field of science, it requires that PC practitioners or collaborators in built heritage field should have high artistic accomplishment and accumulation of history knowledge.

3 APPLYING THE CONCEPT OF PC IN CHINA'S ARCHITECTURAL CONSERVATION

By reviewing the development of PC, it can be seen that its technical method indicated a remarkable rational color, and integrated various advances in the broader discipline of conservation theory and scientific field into its theories and practices. Facing the transformation of the contemporary heritage conservation paradigm, while continuing the past research directions and focuses, how to accomplish related work effectively in more diverse cultural backgrounds is one of the important issues for the PC researchers and practitioners.

The effective use of PC in the multi-cultural context inevitably requires the rational conservation practice and reflection. With reference to an example in China, we attempt to apply the concept of "PC" to China's conservation practices in the form of working packages, to offer several reflections on how to apply PC ideas in diverse policy, management and culture backgrounds.

3.1 *The principles of PC in different cultural backgrounds*

The proposing of the concept PC in built heritage and researches and practices in this field was accompanied by the development of modern conservation thoughts, especially since the ***Venice Charter***. It is in this background that its theories and ethical basics reflect many characteristics of the specific times. Although the birth and development of this concept has an obvious "European gene", the forward-looking features of the "preventive" idea based on a long-term vision and precautionary actions coincide closely with some historical characteristics and inherent requirements for the conservation and inheritance of Chinese architectural heritage.

The scientific characteristics and conceptual connotations of PC discussed above are universal within the different cultural contexts. In other words, there are more common characteristics than differences. As its most important innovations are the way of thinking and the long-term, dynamic vision of planning, the basic ideas and methods, and even specific tools can be adapted and applied to different cultural backgrounds.

3.2 *Main problems when using PC ideas and approaches in different cultural backgrounds*

However, due to the different conservation policies and management methods in different countries, practical problems will also be encountered during the implementation of specific task. Taking China as an example, there are several types of problems in the application of PC:

(1) At present, lack of an operational PC framework with clear steps for guidance, especially in combining these steps with China's existing urban planning systems is one of key problems. The application of PC in China also needs sufficient design in matching the flowchart for the conservation process suggested in the most important cultural heritage conservation guidelines in the ***Principles for the Conservation of Heritage Sites in China***.

(2) Different from other countries' conservation systems, the current highest level of responsible authority of China's built heritage's official conservation and management system includes: National Cultural Heritage Administration (NCHA, former abbreviation SACH) and Ministry of Housing and Urban-Rural Development (MOHURD). Both of them are state-level authority with several Local CH Departments and Local HUD Departments. When conducting a PC project, their executive authority and scope of responsibility need to be clarified. The difference of conservation policies, systems and management structures is a problem that needs to be planned in advance when implementing the idea of PC in different countries.

(3) When specific supporting techniques were introduced, there will be possible controversies and different understandings, due to the difference of building materials and regularity of decay

and the difference in understanding values. Similarly, in terms of community composition and social involvement, there might be many differences with the conservation in the European context.

In China, although the initiatives related to PC appeared somewhat late in government documents, the practices following PC principles have achieved some experience. Although the *Principles for the Conservation of Heritage Sites in China* (the 2004 edition) didn't clearly define PC, it specified the "regular maintenance", "continuous monitoring", "archiving of records" (Art. 20 and Art. 29), and "disaster prevention and preparedness" (Art. 27) as "preventive measures", covering many of the tasks of PC. These measures have been carried out in various conservation projects. For example, during the past decade, experts working for the Imperial Palace have made active explorations on the maintenance, monitoring and risk assessment of the structures and their environment. Other important practices include the monitoring of the Timbered Pagoda of the Ying County, the Main Hall of the Baoguo Temple, etc., which have achieved fruitful results. From 2010 to 2014, Professor Guangya Zhu and Dr. Meiping Wu published a series of studies on PC, systematically introducing its concept and methods to China for the first time (Wu 2014). Generally speaking, there are a lot of monitoring practices and conservation planning in China now, but the innovative inspection work (similar to MOWA model) and the combination of PC ideas and the existing administrative systems are still insufficient. There is a lack of implementation of systematic PC methods within complete working frameworks. Therefore, we tried to make explorations accordingly within the case of Changping.

3.3 Applying the concept and methods of PC in the form of working package- A case study

Changping District is a district of Beijing. It has rich cultural heritage resources. There are three world cultural heritage sites or on tentative list in the region: The Great Wall (Beijing Changping Section), the Grand Canal (Baifuquan Site), and the Imperial Tombs of the Ming and Qing Dynasties (the Thirteen Ming Mausoleums). There are six National Key Cultural Relics Protection Units, three Beijing Cultural Relics Protection Units, and 75 District-level Cultural Relics Protection Units (Figure 5). Many of them are significant ancient buildings or archaeological sites.

During the implementation of PC practice on a regional scale in Changping District, we realize that as the concept and method of PC are universal, its interdisciplinary collaboration, commitment to long-term dynamic planning, and the focus on scientific evidence should be introduced to the professional practice in China and serve as the basic principle in the development of related work. The

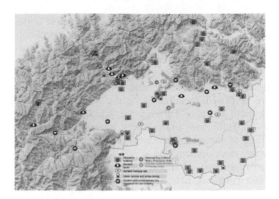

Figure 5. The distribution of cultural protection units in Changping District, Beijing.

main tasks of PC are implemented and integrated, such as regular inspection and maintenance, risk assessment, monitoring, vulnerability index evaluation, etc. Among them, the regular inspection and the risk assessment closely related to the built heritage on the regional scale are rather new in China. Regular inspection has been proved to be very effective in discovering risk in early time and analyzing the material degradation mechanism. Meanwhile, the "working packages" including one or several of these PC tasks are applied and distributed to different users in order to make the achievements separately incorporated into the existing conservation and urban planning systems and management frames, so that PC ideas could be combined with the existing administrative systems.

The vulnerability evaluation form of the regular inspection contains four categories of the first-level catalogue (situations of cultural heritage itself, protection facilities' conditions, environmental conditions, situations of the administration and utilization), eleven categories of the second-level catalogue. The degree of conservation urgency are divided into three levels during the evaluation. The categories of the first-level catalogue and subcategory (second-level catalogue) are as following:

Category A: Situations of cultural heritage itself. The sub-categories include: safety of the main structure, safety of the stylobate and foundation, and safety of the roof.

Category B: Protection facilities' conditions. The sub-categories include: fire control, security facilities, lightning protection, and other protection facilities.

Category C: Environmental conditions. The sub-categories include: environmental conditions inside and beside the structure, environmental conditions of the surroundings, and environmental conditions of the geological carrier.

Category D: Situations of the administration and utilization. The sub-category includes: situations of the administration and utilization (such as illegal constructions, theft, etc.).

In this case study, the team used panoramic aerial photography (Figure 6), 3D laser scanning technology (Figure 7) and non-destructive testing technology (Figure 8) to implement the risk screening to the ancient buildings and ancient sites. During the risk investigation work, the risk assessment indicators are proposed specifically for the different types of cultural heritage. Through the scientific qualitative and quantitative analysis, the comprehensive situation of heritages' security, the status of management, development and utilization, the surrounding environmental changes, geological disasters and other risks, we have tried to establish a reasonable

Figure 6. The panoramic aerial photography in the investigation.

Figure 7. The 3D laser scanning technology in the investigation.

Figure 8. The non-destructive testing technology in the investigation.

and user-friendly evaluation method and indicator system of the risk level, the value status and building condition. Then the system will propose preliminary feedback and comments on risk levels. With expectation, the system will provide basic evaluation data and data files for future preventive conservation, repairment, management, planning, interpretation and other work. During the risk assessment process, the team established a "health condition investigation document" for the cultural heritage which is similar to human's physical examination report. On the one hand, conducting the on-site investigation, survey, measurement, and evaluation to offer the qualitative assessment. On the other hand, the 3D laser scanning, radar imaging technology and high-frequency surface wave imaging technology are practiced on the preliminary non-destructive testing of buildings. In order to carry out a quantitatively assessment, analysis and assessment on diseases and risks such as deformation, inclination, internal fissures, degree of decay and intensity changes have been conducted. It is possible to obtain preliminary data on the ancient building materials, structural characteristics and load-bearing capacity of the built heritage so that the structural problems and potential risks of the building could be noticed as early as possible and take preventive measures in advance.

More importantly, these records are combined with Internet service technology and database technology to store data in the cloud database, directly linked to subsequent regular monitoring, management, tourist navigation, PC measures, sand table, query, update data, etc. The management departments and other related parties can also use these risk investigation records and PC data, panoramic VR data of the cultural heritage on the Internet. In the meanwhile, based on the Internet-based mapping system, a cloud map of built heritage distribution is estab-lished, so that the cloud map publishing template access link can be embedded in the commonly-used life, social network websites and APP (Figure 9) such as Wechat and other mobile communication platforms. This function aims to improve the engagement with the public and encourage the community to participate in and support the PC and daily monitoring of cultural heritage.

This inspection and evaluation work was authorized by the Cultural Committee of Changping District. Within the frameworks of urban planning and cultural heritage administration, the implementation of the risk assessment and other PC tasks and the requirements for subsequent management work are refined in the form of working package modules (Figures 10-11). Since in China, NCHA and MOHURD are respectively responsible for cultural heritage and urban construction related issues, and sometimes their scopes of responsibilities overlap with each other when dealing with issues related to the built heritage. Therefore, different working packages were designed according to their respective scope of responsibility and possible usage modes, so that analysis results achieved in this project can be shared with different administrative departments

Figure 9. The cloud map publishing template of cultural heritage.

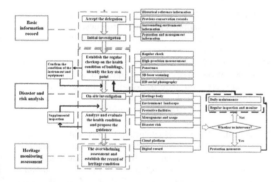

Figure 10. Workflow diagrams.

Figure 11. Fitting into existing urban planning content, the results and requirements of the survey assessment are set up in the form of working package modules.

in the future through these working package modules. For example, the urban overall plan in China of which MOHURD system is in charge usually involves disaster prevention, conservation of National Famous Historical and Cultural Cities, etc. Therefore the achievements of risk assessments can be shared with urban planning departments; NCHA system is responsible for cultural heritage census, preservation and restoration of built heritage, etc. So the damage atlas developed in this case is designed to be shared with

cultural heritage administrative departments. Furthermore, the drawing of risk maps is also one of the current work focuses of NCHA; The local cultural heritage departments are in charge of the conservation status of built heritage, thus working package including regular monitoring and maintenance is set for their needs; and so on (Figure 12). Meanwhile, the users, owners, and managers of each building are identified We share the results of our inspection and recording work with them and help them establish the regular checkup plans on the health condition of the buildings. The data on risk is collected and regularly compared, of which they are also informed (Figure 13-14). During the design for the working package modules, the characteristics of different users are fully considered, in order to make the operations easier for non-experts.

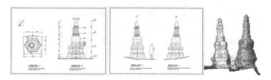

Figure 12. Fitting into existing management system, the results and requirements of the survey assessment are set up in the form of working package modules.

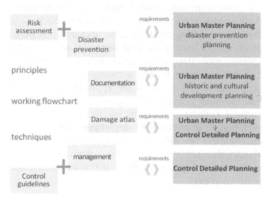

Figure 13. A survey result of a built heritage (recording the angle of inclination of the building).

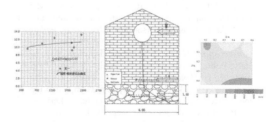

Figure 14. The drawings of regular checkup on the health condition of buildings and the risk investigation plan (the location of cracks and the strength of wall).

By designing user interfaces that are suitable for different users, and packaging our achievements to share with them, we prove that this method is efficient in our cooperation with the heritage related people (building owners, users, visitors, etc.), as well as validity in making their work more scientific.

These working packages, which are easy to understand and operate, ensure the consistency of objectives and methods between multiple participants in the case.

4 CONCLUSION

The scientific characteristics and conceptual connotations of PC discussed above are universal within the different cultural contexts. In other words, there are more common characteristics than differences. As its most important innovation lies in the way of thinking and long-term, circular, dynamic vision of planning, its basic ideas and methods, and even the specific tools developed can be adapted to different cultural backgrounds. The working package modules proposed in this paper is an application to ensure that the main content of PC could work in different cultural context and facilitates the communication with people from different departments of the policy environment in different countries. Moreover, we should also strive to devise more approach to promote the development of PC in different cultural and policy contexts.

ACKNOWLEDGEMENTS

This work was financially supported by the [Scientific Research Foundation of the Graduate School of Southeast University #1] under Grant [number YBJJ1501]; and [Fundamental Research Funds for the Central Universities and the Research Innovation Program for College Graduates of Jiangsu Province #2] under Grant [number KYLX15_0052].

REFERENCES

Brandi, C. 2005. *Theory of restoration (Teoria del restauro)*. Firenze, Italy: Nardini.

Benatti, E., M.P. Borgarino & S. Della Torre. 2014. Planet beni architettonici. Uno strumento per la conservazione programmata del patrimonio storico-architettonico. In S. Della Torre & M.P. Borgarino (ed.), *Proceedings of the International conference preventive and planned conservation, Monza & Mantova, 5-9 May 2014*. Milan, Italy: Politecnico di Milano & Nardini Editore, 13–29.

Chiriac, M., D. Basulto, E. López, J.C. Prieto, J. Castillo & A. Collado. 2013. The MHS system as an active tool for the preventive conservation of cultural heritage. In M. Rogerio-Candelera, M. Lazzari & E. Cano (ed.), *Science and technology for the conservation of cultural heritage - proceedings of the International congress on science and technology for the conservation of cultural heritage, Santiago de Compostela, 2-5 October 2012*. Leiden, The Netherlands: CRC Press/ Balkema, 383–386.

Core, M. 2009. *MDDS: monument damage diagnostic system*. Master thesis, KU Leuven University.

De Rooij, M. 2016. MDCS 2.0 - ondersteuning in het beheer van uw monumenten. In T. G. Nijland (ed.), *Symposium MonumentenKennis "Kennis van de gevel", Amersfoort, December 2016*: 58–65.

Della Torre, S. 2010a. Conservation of built cultural heritage, laws enabling preventive approach: the case of Italy. In M. Guštin & T. Nypan (ed.), *Cultural heritage and legal aspects in Europe*: 169–178. Koper, Slovenia: Institute for Mediterranean Heritage, Institute for Corporation and Public Law, Science and Research Centre, University of Primorska.

Della Torre, S. 2010b. Economics of planned conservation. In M. Mälkki & K. Schmidt-Thomé (ed.), *Integrating aims - built heritage in social and economic development*: 143–157. Espoo, Finland: Centre for Urban and Regional Studies Publications, School of Science and Technology, Aalto University.

Della Torre, S. 2013. Planned conservation and local development processes: the key role of intellectual capital. In K. Van Balen & A. Vandesande (ed.), *Reflections on preventive conservation, maintenance and monitoring of monuments and sites*. Leuven, Belgium: ACCO, 123–127.

Ferreira, T.C. 2018. Bridging planned conservation and community empowerment: Portuguese case studies. *Journal of Cultural Heritage Management and Sustainable Development* 8(2): 179–193.

Haagenrud, S.E., J.F. Henriksen, J. Veit & B. Eriksson. 1998. Wood-assess: systems and methods for assessing the conservation state of wooden cultural buildings. In *Materials and technologies for sustainable construction, Symposium A, CIB World Building Congress, Gävle, Sweden, 7-12 June 1998*.

Henriksen, J.F., S.E. Haagenrud, U. Elvedal, J. Häusler, P. Norberg & J. Veit. 1998. Wood-assess - mapping environmental risk factors on the macro local and micro scale. In *Materials and technologies for sustainable construction, Symposium A, CIB World Building Congress, Gävle, Sweden, 7-12 June 1998*.

Ibáñez, A.J.P., J.M.M. Bernal, M.J.C. De Diego & F.J. A. Sánchez. 2016. Expert system for predicting buildings service life under ISO 31000 standard. Application in architectural heritage. *Journal of Cultural Heritage* 18: 209–218.

ICOM-CC. 2008. Terminology to characterize the conservation of tangible cultural heritage. http://www.icom-cc.org/242/about/terminology-for-conservation/#.XAYjWK5fj3g, 2018-12-01.

Muñoz Viñas, S. 2005. *Contemporary theory of conservation*. Amsterdam, Netherlands: Elsevier.

Stulens, A., V. Meul & N. Čebron Lipovec. 2012. Heritage recording and information management as a tool for preventive conservation, maintenance, and monitoring: the approach of Monumentenwacht in the Flemish region (Belgium). *Change over Time* 2(1): 58–76.

Van Balen, K. 2015. Preventive conservation of historic buildings. *Restoration of Buildings and Monuments* 21 (2-3): 99–104.

Vandesande, A. 2017. *Preventive conservation strategy for built heritage aimed at sustainable management* and local development. Ph.D. thesis, KU Leuven University.

Vandesande, A., K. Van Balen, S. Della Torre & F. Carodoso. 2018. Preventive and planned conservation as a new management approach for built heritage: from a physical health check to empowering communities and activating (lost) traditions for local sustainable development. *Journal of Cultural Heritage Management and Sustainable Development* 8(2): 78–81.

Wu, M. 2014. *Preventive conservation of architectural heritage in China*. Nanjing, China: Southeast University Press.

Zalamea Patiño, O., J. Van Orshoven & T. Steenberghen. 2016. From a CityGML to an ontology-based approach to support preventive conservation of built cultural heritage. In *Proceedings of the 19th AGILE International Conference on Geographic Information Science, Helsinki, 14-17 June 2016*.

Preventive Conservation - From Climate and Damage Monitoring to a Systemic and Integrated Approach – Vandesande, Verstrynge & Van Balen (eds)
© 2020 Taylor & Francis Group, London, ISBN 978-0-367-43548-6

Preventive and planned conservation: Potentialities and criticalities, strategy and tools, lessons learned

Rosella Moioli
Professional architect

ABSTRACT: The paper deals with a professional experience of Preventive and Planned Conservation (PPC) applied to the Built Cultural Heritage in the framework of the strategy conceived in Lombardy Region. Since 2003 there have been many field trials of implementation of Conservation Plans (CP) for historic buildings, it could be of some interest to try an analysis about the evolution of the tools and of the different impacts depending on the characteristics of the owner, the regulation frameworks and constraints, and on professionals/technicians' competences. The CP is based on technical standards for implementation developed under the theoretical point of view, considering knowledge as the fundamental premise, while the choice of digital instruments has been an open issue due to the high speed of the digital innovation. Furthermore, the paper will investigate the relationship among effectiveness of CP and digital technologies for the management of information, people awareness, the economic convenience of PPC.

1 INTRODUCTION

The paper deals with 20 years of professional experience on Preventive and Planned Conservation (PPC) applied to the Built Cultural Heritage in the framework of the strategy conceived in Lombardy and promoted by the Regional Government and Cariplo Foundation.

It is a reflection both on the theoretical approach, and its evolution, and the on the field experiences: since 2003 there have been many field trials of implementation of Conservation Plans for historic buildings, thus it could be of some interest to try a qualitative analysis about the evolution of the tools and of the different impacts depending on the characteristics of the owner, the regulation frameworks and constraints, and on professionals/technicians' competences. I had the chance to be involved, in different roles, in most of the actions promoted by Lombardy Region and then by Cariplo Foundation for the dissemination in Lombardy of a processual long-term approach to conservation.

Moreover, as a professional architect, I got the opportunity to carry out some Conservation Plans whether in case of restoration designs or also for the planning of the conservation activities for the management of the sites.

The paper has therefore the ambition to give a contribution to the research on preventive conservation through the reflections carried out as a participant observer of the process. The methodology of participant observation is widely adopted in researches aimed about people, behaviours, processes and cultures (Kawulich 2005), and the aim of the paper is to investigate issues like the effectiveness of Conservation Plans and digital technologies in supporting behavioural changes, people awareness about heritage values, the perception of economic convenience of Preventive and Planned Conservation. These topics can hardly be dealt with implementing quantitative methods, as the most important targets of the action are often secondary effects, or even externalities, which are hidden behind figures, so that their detection requires the involvement of actors in sight of qualitative assessments, which can be correctly produced by the participant observer.

The paper will firstly analyse the process, which led to the definition and the theoretical framework of "conservazione programmata" (Planned Conservation) as carried out in Lombardy for two decades.

Then the main tool of the strategy is introduced, i.e. the Conservation Plan (CP), also in its evolution towards digitalization. Reflections about the role of the operational tool have been developed both taking into account the course of the research and the hands-on implications of its implementation. The main remark is that the full effectiveness of the CP can be achieved only with the stakeholders' awareness of the whole Conservation process.

The third section deals with the on-field experiences, describing and comparing the different perspectives of the calls issued by Cariplo Foundation, directly targeted on preventive conservation or framing the strategy into the broader concept of comprehensive wide-area projects, and of professional experiences.

The comparison leads to a series of remarks, presented in the fourth section of the paper.

2 PREVENTIVE AND PLANNED CONSERVATION IN THE ITALIAN FRAMEWORK

2.1 *Definition*

Preventive and Planned Conservation, according to the Italian framework, is a long-term strategy aiming at the integration of conservation and valorisation activities for an effective management of Built Cultural Heritage. It is oriented to prevention and continuous care and it is a process aimed at the production of knowledge that needs tools for the management of information, and for the planning of the conservation activities.

This short but meaningful definition is the result of twenty years of reflections, researches and hands-on trials on the topic of prevention and planning of the conservation activities applied to the Built Cultural Heritage. They have been carried out mainly in the Lombardy region, thanks to the cooperation of several stakeholders such as Lombardy Region Government, Politecnico di Milano, the Central Institute for Restoration, now ISCR, and Cariplo Foundation. The strategy has been developed in time and it has been transformed and improved by a continuous work of research and on the field applications, besides being enriched by exchanges and discussion with the main actors of the international debate.

The expression Preventive and Planned Conservation originated in the specific Italian context, where preventive and planned are the key words for bridging the new research with Brandi's (Brandi 2005) and Urbani's (Urbani 2000) theories. Furthermore, "planned" plays an important role in the long-term vision typical of the Conservation process.

Although this strategy has been conceived in a national context, the researchers have been aware of the importance of the international debate and of the necessary dialogue with those organizations which were (and are) involved in similar studies and activities. The first attention has been given to the Monumentenwacht organization and then to the Raymond Lemaire International Centre for Conservation, and consequently to the UNESCO Chair on Preventive Conservation, monitoring and maintenance of monuments and sites. The goals are similar, and many contacts points have been found. The adjective "planned" in the Italian research would like to stress the need for a managerial attitude in the field of Built Cultural Heritage.

A second important reference is the Preventive Conservation praxis applied to the movable objects and to the historic interiors (Staniforth 2013). This kind of approach is put into practice with some specific and standardized procedures (Forleo et al. 2017). Thus, Preventive Conservation for a long time has been commonly meant as a practice for the conservation of art works, objects and precious surfaces. Now the aim is to make evident the relational nexus among the building, the interiors and the objects contained within.

As a premise, in this paper PPC will be meant in a holistic way referring to the oneness of the conservation process, in agreement with the Italian law for the protection of Cultural Heritage (Della Torre 2010/a). Then it is important to clarify a terminological and theoretical issue: "restauro", i.e. the direct intervention on historic things, is understood as one of the activities included in the more extensive notion of conservation process. In the paper "restauro" will be translated in "restoration", but it should be clear that in this context the meaning of restoration doesn't necessarily include the revival of "the original concept or legibility of the object" (Feilden 1982).

2.2 *The theoretical framework*

The origin of the researches in Lombardy was the "pilot project for the Umbria region" launched in 1976 by Giovanni Urbani, the first scholar who introduced the concept of planned maintenance, or planned conservation, as once in his writings he referred to (Minosi 2005).

With these premises and thanks to a theoretical reasoning about the fundamental notions of conservation, preventive effectiveness, maintenance and planning it has been possible to define the new approach for a PPC strategy, combining a strong theoretical basis with an operational comparison with the regulations for public works and for the protection of Cultural Heritage.

The main, only in appearance nominalistic, topic of our reflection has been the language switch from the word "maintenance" to the word "conservation" for it is representative of a specific approach to the Built Cultural Heritage: the building is not considered as the sum of technological elements which must reach standard levels of performances, as foreseen by the praxis of planned maintenance in the facility management, with the consequently substitution of the not efficient elements. Quite the opposite, the conservative approach considers the real behaviour of each element as the starting point for a process made of checks, monitoring, preventive action and maintenance in order to maintain as long as possible their material authenticity. Thus, it means that the conservation approach pays attention to the vulnerability of fragile elements and materials, with the awareness that often these characteristics actually are the values, tangible and intangible, to preserve. This kind of reasoning, in an early stage, gave rise to the Italian debate since the 1980s with the theories of conservation by Marco Dezzi Bardeschi (Dezzi Bardeschi 1991) and Amedeo Bellini (Bellini 1999) which, in an extreme synthesis, state that it is not possible to make a selection of a specific conservation state, assuming it can be maintained in an ever-ending equilibrium state, and in this way the only choice is to tackle the action of passing time and managing the unavoidable transformation.

The progress of these considerations contributed to the cultural background which generate the wording of the Art. 29 of the Italian law for the protection of Cultural Heritage, that is D.Lgs. 42/2004. It defines Conservation as the result of "a coherent, coordinated and planned activity of investigation, prevention, maintenance and restoration". More recently these reflections led to the idea that it is no longer possible to consider the building as a static object, but on the contrary it is a dynamic system. Hence, conservation activities must aim at recognizing the building's peculiarities and its potentialities of evolution (Della Torre 1999) or even better the co-evolution potentialities (Della Torre 2010/b; Della Torre 2019).

All these arguments have been developed in the last years also through a frequent dialogue of Italian researchers with the UNESCO Chair on Preventive Conservation in Leuven (Van Balen & Vandesande 2013; Van Balen 2015; Vandesande 2017), definitely useful to clarify the concepts, and also some differences due to the cultural backgrounds, contexts and legal frameworks.

2.3 The conservation plan

PPC is a long-term process which needs operational tools in order to be carried out and applied in the different contexts and cases. If PPC is a strategy, it is necessary to provide a tool for its implementation: such tool is the Conservation Plan (CP), structured according to the guidelines of Lombardy Region Government, adopted in 2005, and based on technical standards (Della Torre 2003). These standards have been developed mainly under the conceptual and theoretical point of view, considering knowledge as the fundamental premise, while the choice of digital instruments for the compiling of the technical documents has been an open issue due to the high speed of the digital innovation (Della Torre, Moioli & Pili 2018).

The CP encompasses four documents: Technical Handbook, Conservation Program, Economic Budget and User Handbook.

The Technical Handbook defines the framework of knowledge and previous events that concerned the building through the description of materials and construction techniques, decays, activities and technical operations, diagnostics, relationship between the elements and risk assessment. The aim is to pinpoint the necessary preventive actions. These activities consist in inspections and direct and indirect actions on the building, according to the given definitions of prevention and maintenance by the Italian law for the protection of Cultural Heritage.

The Conservation Program is connected to the Technical Handbook and defines by whom, when, and how the activities have to be carried out.

Subsequently, the activities are programmed and systematized into a time schedule.

The Economic Budget systematizes the economic resources management through definition and planning of costs.

The User Handbook is the document that summarizes in a non-technical language the information contained in the Technical Handbook and in the Conservation Program, targeting end-users in order to pursue conservation through continuous care and proper management. The User Handbook expresses and highlights the active role of the user in the conservative strategies, as a guarantor of good conservation, being he/she who daily lives the building. (Della Torre, Moioli & Pili 2018)

The CP is the tool for the implementation of the strategy, and it is characterized, according to the proposed model, by flexibility from the point of view both of the format and of the digital instruments needed for its filling in.

The first tests aimed at understanding the main features and shaping the CP. This choice led to test different digital solutions: simple word processing programs, a first attempt of data-base software; spread-sheets and, later on, a specific Informative System (Benatti, Borgarino & Della Torre 2014; Della Torre, Moioli & Pili 2018).

The opportunity of testing the methodology in different situations highlighted the potentialities of PPC approach, and of the different operational tools used to draw up the plans, but also some important criticalities. These tests in fact have been very effective in order to verify the limits of the regional guidelines and to improve the conceptual structure of the CP. At the same time a number of difficulties came to light, related to the not adequate procedures of public bodies for the planning of maintenance works and to the lack of competences on Built Cultural Heritage.

It has been clear that the major difficulties were not related to the complexity of the technical documents of the CP or to the "digital ignorance", but rather to a general resistance to change some practices in the sector of public works and, also, in the management of private heritage properties. Even if this change would have been a factor of economic savings, of time savings in the full swing and much more effective for the collection and systematization of the information and data for future costs calculation, statistics and knowledge.

It is possible to detect a twofold origin: on one hand, a general attitude which affect public sector, and which is not our interest in this paper, and on the other hand a lack of specific competences in the field of Built Cultural Heritage due to the fact that the qualification of professionals and of public officers is not mandatory (Van Balen & Vandesande 2019).

Having observed this kind of response, it has been necessary to shift to a more methodological level and try to involve the stakeholders in the design of strategies and not only in the drawing up of the technical documents.

It was clear from the beginning that the proposal was not referred simply to the field of facility management, better to say of maintenance or preventive conservation activities, but it was a strategy for the management of all the activities related to the building: conservation, valorisation, communication and enjoyment.

3 ON THE FIELD EXPERIENCES

3.1 Research promoted by Lombardy Region (2000-2006)

The researches financed by Lombardy Region on the topic of PPC started in 1998, and the formal approval, in 2005, of the regional "guidelines for the experimentation of Planned Conservation methodology for the historic buildings" has been the conclusion of this first step of the long path.

The institution of a regional centre for the Risk map of Cultural Heritage has been the initial switchover (Cannada Bartoli, Palazzo & Urbisci 2003). The Risk map is the broader national system of databases able to collect information about potential hazards affecting cultural heritage in relation with the context.

The first experiences can be divided in two phases. In the very beginning the questions concerned the analysis of the processes and the definition of tools capable to trigger the foreseen changes. Once the maintenance plan had been identified as the crucial point, the research focus moved on how to enhance the maintenance plan in order to set up adequate practices for historic buildings, keeping unicity and material authenticity. As referred above, the reflection on the risks of an unthinkingly maintenance qualified the outputs of these phase, and by this step new and quite surprising bridges were built among the sectors of historic preservation and facility management. This phase ended up in the presentation of a book (Della Torre 2003), in which the theoretical and methodological background of all the contents of the "conservation plan" are discussed, and the relevant guidelines are presented. Perhaps, the most advanced proposal was to identify the CP and the scientific report, or as-built documentation, of restoration works: a wishful requirement of the Italian law on public works concerning protected historic buildings, usually overlooked in the reality. The background of this identification was the idea of turning the CP into an information system, to be prepared before the works and updated after, in order to plan maintenance activities on the basis of a detailed and up-to-date knowledge of the buildings and its issues.

The second phase was focused on the development of electronic tools in order to put this idea into practice. In those years Lombardy Region had a strong commitment on developing the Risk Map and creating an information system for cultural heritage (Cannada Bartoli, Palazzo & Urbisci 2003; Cannada Bartoli & Della Torre 2005; Cannada Bartoli & Della Torre 2013). The research on Risk Map was the cuddle for planned conservation research, and also the proposal of giving the regional system new tools for new practices was perfectly aligned with the vision carried out by the general director for Culture, Pietro Petraroia, also for the perspective of a future connection between the multiple levels of the processes. Undoubtedly, everyday practices registered in the as-built reports and in the maintenance plans produce a lot of data, which could feed public data-banks; on the other hand, public data-banks can provide all the operators with fundamental data, e.g. on territorial risks, giving a strong input for higher quality.

The output of these research was a software called SIRCoP (Regional Information System for Planned Conservation), which was the object of several courses and tests.

Those first experiences were fundamental to the definition of a vision and a set of tools, which in the following years had to be implemented, and slowly understood by the regional system. In the reality, the activities had been applied, that is carried out on real tasks and discussing with the involved stakeholders, but the real target still was the development of the procedures and the tools.

3.2 Cariplo foundation grants (2006 onwards)

Fondazione Cariplo, the most munificent banking foundation in Italy and one of the most important charities in the world both for assets and grants, since its establishment in 1991 by statute supplied grants to support cultural heritage conservation (Cammelli 2007), and since 2008 financed projects for the fostering of PPC. The aim was to support projects based on the principles of continuous care, production of new knowledge, integration of conservation and valorization activities in sight of a real local development.

A first important action has been the integration of PPC within the regional plan "Cultural Districts".

Then a specific call has been launched and repeated for ten years: a grants program for projects aiming at the dissemination of technologies, methodologies and good practices concerning the PPC. Through the years the calls got diverse titles and were issued on different lines, because the grants have been thought and updated thanks to a productive dialogue with the territories and local stakeholders.

3.3 Cultural districts

In order to face the constant and growing demand of restorations, felt as mandatory by communities and local administrations, and to answer at the same time the above described need to innovate the processes, for the last fifteen years Cariplo Foundation matured an approach, which aims at privileging projects more organic and aware than the mere restoration, asking the projects to include the sustainable reuse and valorisation of Built Cultural Heritage and the implementation of PPC. The Cultural Districts plan, promoted in 2005 as a pre-feasibility study, has been a first wide experimentation of this approach. In five years, through progressive steps of selection, involvement and accompaniment, the result has been the real implementation of six projects on their own territories, in the years 2010-2014.

The Cultural Districts plan was born with the target to promote an integrated cultural planning, looking for projects, in which heritage conservation activities could play a central role in the empowerment of all the involved actors, in the enhancement and valorisation of skills and intellectual capital and in the integration of diverse supply chains (Barbetta, Cammelli & Della Torre 2013). Through the years, by monitoring the impacts of the actions and elaborating a continuous reflection, Cariplo Foundation took a more and more active role of motivation and orientation. The financial dimensions of the plan cannot be overlooked: Cariplo Foundation granted the projects with the sum of more than 20 million of Euros, matching investments by the beneficiaries for more than 32 million.

Besides the traditional co-financing leverage, to guarantee the commitment by beneficiaries, the calls have been designed with an increasing care in order to hit non only the basic targets, such as in the beginning the accomplishment of restorations, but also methodological aims, in the field of innovation and capacity building, so that a real development of the territories could happen, and in all the sectors.

In the specific case, it is worthy to underline that Cariplo Foundation, by issuing these calls, de-scribed as "grants under condition", decided to take an active role in the complex debate on the role of cultural heritage in the various local sectors. Up today, the outcomes of this outstanding experience (CHcfE Consortium 2015: 197) has been discussed mainly under the perspective of the impacts on policy-making or on innovation and local development (e.g. see Fanzini & Rotaru 2012; Cerquetti & Ferrara 2015; Della Torre 2015), sometimes even under evaluating or misunderstanding the differences between the many Italian projects labelled as "distretto culturale" (Ponzini, Gugu & Oppio 2014; Nuccio & Ponzini 2017).

The strategy chosen by Fondazione Cariplo had Built Cultural Heritage as a pivot of the comprehensive vision. For this reason, a special importance has to be acknowledged to the evaluation of the impacts of conservation activities, understood not only under the narrow perspective of the restoration sector, but in the wider and multidisciplinary frame of an integrated valorisation. The methodology for these evaluations passes through the involvement of participants in the evaluation reflections, turning evaluation into and awareness exercise (Della Torre & Moioli 2012).

3.4 Grant program for preventive and planned conservation

In strict consistency with the workflow of the Districts, all of them including a more or less strong strategic line PPC, in 2008 Cariplo Foundation for the first time promoted a line devoted to support "Planned Conservation": in that moment the choice was definitely innovative on behalf of a charity and still today it is pretty timely for its purpose of making prevention, based on knowledge, recognized as an essential part of the overall management of a cultural property. This choice induced organizations to optimize the use of resources and to program on the medium-long run.

Between 2008 and 2016 Cariplo Foundation promoted, through these calls, an innovative role of cultural heritage in the local development mechanisms, supporting actions based on a long-term strategic vision, capable to put into practice an important innovation of process: the step from restoration meant as a one-time thing, to continuous care of the goods.

The structural shortage of financial support is a problem that should be brought to decision makers' attention. In the reality, the incentive provided by Cariplo's call triggered the maybe opportunistic adoption of the proposed model. But as it is sure that prevention, even if not so appreciate in the private perspective also because of the short term of the evaluations, is convenient in a macroeconomic perspective on the long run, other kinds of incentive would be desirable. For instance, the plain reduction of the added value tax, which at least in Italy apply on diagnostic and maintenance activities discouraging rates in comparison with new construction or radical intervention.

The decision to applicate to the Call seldom came from the owners, but initiatives were promoted by other subjects, such as universities or specialized professionals, who were informed of the opportunity. Some promoters were definitely not updated on the methodology. Only in few cases the guidelines promoted by Lombardy Region and the related Information system (SIRCoP) have been implemented (nor did Fondazione Cariplo require the implementation).

As the Call was reserved to public bodies (or similar), the projects went through all the problems related to regulations not oriented to long-term activities, and to the lack of sensitivity and skills in the public sector (Moioli & Baldioli 2018).

The experiences financed by Cariplo in the frame of the specific calls have been interesting also for the richness of the funded contents feeding the conservation plans, as the criteria of the call were set to promote high scientific quality and multidisciplinary cooperation. Therefore, the teamwork on cases like Palazzo Besta in Teglio (Foppoli et al. 2012), the Monastery of Santa Maria del Lavello (Daniotti et al. 2014: Erba et al. 2016) and the churches of Vimercate (Moioli 2015) enabled to learn several lessons on what the implementation of a holistic approach to preventive conservation can mean, and also to observe biases, barriers and criticalities.

3.5 *Professional experiences*

The policies supported by the Regional Government and the incentives provided by Cariplo Foundation entailed in Lombardy region a climate, in which the drafting of a CP was often required. For a professional involved in the researches and the on-field tests, this could be a competitive advantage, if the decision makers were able to understand the convenience of sharing and supporting this innovation.

Therefore, another scenario for the implementation of conservation plans has been offered by a set of major restoration works, financed by public bodies with the support of competitive calls, in which the presence of an advanced maintenance plan started to be appreciated as one of the preference criteria.

Already in the first years from the 21st century, I had the opportunity to experiment the drafting of the maintenance plan during the design of the restoration of Villa Sottocasa in Vimercate (Moioli 2004), the Hall of the Fame in the Castle of Bellusco (Moioli 2009), developing a set of remarks on the implementation of maintenance planning already in the design phase. In other cases, the conservation plan was compiled ex-post in cooperation with other professionals, as happened for three buildings in Valtellina, with the support of Interreg funds (Bossi, Calegari & Moioli 2016).

In terms of Participant Observation, these experiences have been invaluable in order to understand the different points of view of the professionals invited to change their habits and learn new processes.

4 REMARKS

The main remarks, which are worthy noting concern the relationships among the effectiveness of CP and: 1) the perception of economic convenience of PPC; 2) people awareness; 3) different management models.

4.1 *Economic convenience*

PPC needs a long-term perspective. The strategy requires some years to show its efficacy in terms of positive economic impacts, as well as of better conservation of heritage goods thanks to prevention and maintenance activities.

The objections raised by who doubts about the economic convenience of the continuous care have some basis, if the current practices keep being considered the only conservation activity. Instead, the whole conservation process should be considered, and integrated with a valorisation not to be meant as commodification and exploitation of heritage (Van Balen & Vandesande 2016). In short, those objections hold only if the returns of investments in conservation are evaluated just in financial terms and concentrated in the very short time of restoration.

Therefore, the first answer to scepticism should be a management of cultural heritage focused on long term processes, capable to merge conservation and valorisation activities on the perspective of an ordinary, even daily management of the properties, and to consider the making of the positive externalities produced.

The main externalities observed in the carried out experiences of the PPC Strategy are: an enhanced attention paid to the production of knowledge (new information on heritage, on traditional materials and techniques…); preservation of the set of values associated to goods (cultural, social, artistic…); development of intellectual capital (capacity building for the involved subjects, entailing a future reduction of the costs due to incorrect and harmful practices and interventions); development of social/relational capital (strengthening of territorial relationships and identity).

The recognition of positive externalities, however, introduces the theme of an evaluation, which should not be only monetary, nor limited to direct impacts. From the beginning, one of the targets has been to highlight the non-monetary benefits for the economic actors, including construction enterprises and the building sector at large. The topics of training specialized craftsmen, of creating a market niche and developing innovative techniques and intervention procedures have surely been a fil rouge that links all the researches and experimentations since the early 2000s. In some of the granted projects these themes emerged very clearly.

4.2 *Awareness*

Care is a consequence of the shared recognition of value, it's the output of processes, which are also processes of social cohesion. In many experiences, even worldwide, this happened through projects entailing first-hand mobilization and action, playing on the recovery of simple empiric knowledge and direct action (Garcia, Cardoso & Van Balen 2015), carried out by ordinary people in strict cooperation with experts and technicians. These experiences have a high social value, they work definitely well in front of buildings made up by traditional and local technologies, but they cannot be exploited too far when dealing with layered and fragile historic buildings, which require high professionalism, besides training for safety.

Community involvement is anyway a mandatory condition to make changes durable, or fully sustainable, but changes cannot happen without the presence of the experts. Instead, the role of experts has to be redefined: no longer authoritative, but maieutic, careful to the needs and the potentialities expressed by fragile communities, which sometimes are even still being formed.

Cultural heritage, once upon a time prerogative of the elites, has got today, in the society of the even chaotic access to information, mainly a function of social connector and activator. This function, after the first step of awareness, becomes more effective just through the participation to the processes of PPC.

One of the main targets has been to guide the property managers towards a strategy that in the future could avoid that restorations take huge costs because of the lack of care and maintenance, or because of the low quality of the works. Thus, change of attitude could happen in medium-long terms and will require that all the involved actors (managers, public officers, professionals, contractors and conserver-restorers) enhance their skills.

In the directly observed experiences, the problem of guaranteeing a high quality was often detected, as if the importance and the value of quality were misunderstood, and the related costs were not seen as investments. This observation confirms that the concept of quality requires a broader understanding, at the crossroad of technical knowledge, stakeholders' involvement and update of the procedures (Van Roy 2019).

As a matter of fact, these kind of problems seem to have roots far behind, that is in the lack of sensitivity of many professional for historic preservation: which is legitimate, but then it is at least strange that this architects or technicians without competences work on listed buildings, if they have no skills nor interest for a sector, which requires a special professionalism (Van Balen & Vandesande 2019). Furthermore, basic information is not shared: many professionals ignore that since 2004 the Italian framework preservation law introduced a processual vision of the conservation process, including prevention and maintenance. Instead, the specialized professionals tend to resist against what they see as new duties, never paid enough, but in general they can understand the reasons and the contents of planned conservation procedures. In the reality, the empowerment of officers, professionals and contractors through educational courses proved to work better, as the attendees were selected on a voluntary basis.

4.3 Management models

The management model is a strategic tool for an effective PPC approach integrated with the valorisation activities. The issues of daily management respond to the requirements of efficiency and efficacy of the investments, as a continuous control of the historic building and the relevant activities is the basis for the evaluation of the concrete effects of the strategies.

The transition to planned strategies proved to require relevant change management policies, implementing procedures at different levels in an integrated way, as an effective management of cultural heritage properties has poor traditions in Italy.

5 CONCLUSIONS

The possibility to turn into reality the approach proposed by PPC depends on several factors, such as the demonstration of its convenience, the specialization of the competences of technicians, the adjustment of the regulatory system.

Then, PPC, relaunched by Fondazione Cariplo calls, posed and poses multiple challenges to owners, managers and technicians: to implement a long-term vision and an attitude to plan activities; to make the decision makers willing to spend for not so visible actions, sometimes just preparatory to something else, some-times focused on registering information not appreciable soon; to understand the utility of investments in surveying, and diagnostics and monitoring, seemingly not decisive, instead of choosing easy (but not durable) solutions; to learn more advanced techniques and to understand the importance of quality, therefore accepting the related costs; to understand the importance of continuous attention and knowledge management, therefore getting prepared for regular inspections and the implementation of information systems, in which to file available data and data under production.

As a provisional conclusion, the proposal emerges to set up associate centres for the supply of services, looking at the Monumentenwacht models and at the experiences successfully tested by Distretti Culturali projects. In these examples services for preventive planned conservation could be supplied on the basis of the lesson learned in Monumentenwacht experiences. This means that stakeholders should be treated not just as clients but as involved actors, and services should include knowledge enhancement as a precondition for sustainability and the enhancement of quality (Vandesande 2017; Van Roy 2018).

REFERENCES

Barbetta G.P., Cammelli, M. & Della Torre, S. (eds) 2013. *Distretti culturali: dalla teoria alla pratica*. Bologna: Il Mulino.

Bellini, A. 1999. De la restauración a la conservación; de la estética a la ética. *Loggia: Arquitectura y restauración* 9: 10–15.

Benatti, E., Borgarino, M.P. & Della Torre, S. 2014. Planet Beni Architettonici. Uno strumento per la Conservazione Programmata del Patrimonio storico–architettonico. In S. Della Torre (ed.), *ICT per il miglioramento del processo conservativo*: 13–29. Florence: Nardini.

Brandi, C. 2005. *Theory of Restoration*. Florence: Nardini.

Caligari, A., Bossi S. & Moioli, R. 2016. A sustainable management process through the preventive and planned conservation methodology: the conservation plan of the complex of S. Antony. In R. Amoeda, S. Lira & C. Pinheiro (eds), *Heritage 2016. Proceedings of the 5th International Conference on Heritage and sustainable development*. Barcelos: Green Lines Institute for Sustainable Development.

Cammelli, M., 2007. Beni culturali: conservazione e valorizzazione. Le Fondazioni di origine bancaria e il restauro dei beni culturali, *Aedon – Rivista di arti e diritto on line*, n. 2.

Cannada Bartoli, N., Palazzo, M. & Urbisci S., 2003, Carta del rischio del patrimonio culturale. Il polo regionale della Lombardia, *Bollettino ICR* 6-7: 4–25.

Cannada Bartoli, N. & Della Torre, S. 2005. Programmare la Conservazione: verso un Sistema integrato di documentazione dei Beni Culturali in Lombardia. In P. Croveri & O. Chiantore (eds), *Monitoriaggio del Patrimonio Monumentale e Conservazione Programmata*: 8–17. Florence: Nardini.

Cannada Bartoli, N. & Della Torre, S. 2013. Verso la conservazione programmata, *Rivista dell'Istituto per la storia dell'arte lombarda* 9: 7–18.

Cerquetti, M. & Ferrara, C. (2015). Distretti culturali: percorsi evolutivi e azioni di policy a confronto. *Il Capitale culturale, Studies on the Value of Cultural Heritage*, 3: 137–163. Available at: http://riviste.unimc.it/index.php/cap-cult.

CHCfE Consortium. 2015. *Cultural Heritage Counts for Europe, full report*. Retrieved from: http://bit.ly/2jERIwx.

Daniotti, B., Erba, S., Rosina, E., Sansonetti, A. & Moioli R. 2014, PPC at Lavello convent: towards a sustainable conservation plan after the restoration. In Della Torre, S. (ed), *Metodi e strumenti per la prevenzione e la manutenzione*:137–150. Florence: Nardini.

Dann, N. 2004. Owners' attitude to maintenance, *Context* 83: 14–16.

Della Torre, S. 1999. "Manutenzione o "Conservazione"? La sfida del passaggio dall'equilibrio al divenire. In G. Biscontin & G. Driussi (eds), *Ripensare alla manutenzione*: 71–80. Venice: Arcadia Ricerche.

Della Torre, S. (ed) 2003. *La conservazione programmata del patrimonio storico-architettonico: linee guida per il piano di manutenzione e il consuntivo scientifico*. Milano: Guerini.

Della Torre, S. 2010/a. Conservation of built cultural heritage, laws enabling preventive approach: the case of Italy, in M. Gustin & T. Nypan (eds), *Cultural Heritage and legal Aspects in Europe*: 168–178. Koper: Institute for Mediterranean Heritage, Institute for Corporation and Public Law, Science and Research Centre, University of Primorska.

Della Torre, S. 2010/b. Preventiva, Integrata, Programmata: le Logiche Coevolutive della Conservazione. In: G. Biscontin & G. Driussi (eds), *Pensare la Prevenzione: Manufatti, Usi, Ambienti*: 67–76. Venice: Arcadia Ricerche.

Della Torre, S. 2015. Shaping tools for Built Heritage conservation: from architectural design to program and management. Learning from "Distretti Culturali". In K. Van Balen & A. Vandesande (eds), *Community involvement in heritage*: 93–101. Antwerp: Garant.

Della Torre, S. 2018. The management process for built cultural heritage: Preventive systems and decision making. In K. Van Balen & A. Vandesande (eds), *Innovative Built Heritage Models*: 13–20. Leiden: CRC Press/Balkema.

Della Torre, S. 2019. A coevolutionary approach to the reuse of built cultural heritage. In G. Biscontin & G. Driusssi (eds), *Il Patrimonio Culturale in mutamento. Le sfide dell'uso*: 25–34. Venice: Arcadia Ricerche.

Della Torre, S. & Moioli R. 2012. Designing an active monitoring system: the Planned Conservation Project in Monza and Brianza Province. In S. Mendes Zancheti & K. Similä (eds), *Measuring Heritage Conservation Performance*: 142–147. Olinda & Roma: CECI & ICCROM.

Della Torre, S., Moioli, R. & A. Pili, A. 2018. Digital tools supporting conservation and management of built cultural heritage. In K. Van Balen & A. Vandesande (eds), *Innovative Built Heritage Models*:101–106. Leiden: CRC Press/Balkema.

Dezzi Bardeschi, M. 1991. *Restauro: punto e da capo. Frammenti per una (impossibile) teoria*. Milano: Franco Angeli.

Erba, S., Moioli, R., Sansonetti, A., Rosina, E. & Suardi, G. 2016. From tradition to innovation: plaster at risk under severe climatic conditions. In K. Van Balen & Verstrynge, E. (eds) 2016. *Structural Analysis of Historical Constructions – Anamnesis, diagnosis, therapy, controls*: 185–192. London: Taylor & Francis Group.

Fanzini, D. & Rotaru, I. 2012. The Italian Cultural District as a model for sustainable tourism and territorial development. *Journal of Tourism Challenges and Trends*, 2(2): 11–34.

Fielden, B. 2003. *Conservation of Historic Buildings*. Third Edition. Oxford/Burlington: Architectural Press.

Foppoli, D., Realini, M., Colombo, C. & Moioli R. 2012. Le facciate dipinte di Palazzo Besta (Teglio). Valutazione e gestione del rischio, in G. Biscontin & G. Driussi (eds), *La conservazione del patrimonio architettonico all'aperto. Superfici, strutture, finiture e contesti*: 821–831. Venice: Arcadia Ricerche.

Forleo, D., De Blasi, S., Francaviglia, N. & Pawlak, A. 2017. *EPICO – European Protocol In Preventive Conservation, phase 1. Methods for conservation assessment of collections in historic houses*. Genoa: Sagep Editori.

Garcia, G., Cardoso, F. & Van Balen, K. 2015. The Challenges of Preventive Conservation Theory Applied to Susudel, Ecuador. In K. Van Balen & A. Vandesande (eds) 2015. *Community involvement in heritage*: 117–130. Antwerp-Appeldoorn: Garant.

Kawulich, B.B. 2005. Participant Observation as a Data Collection Method. *Forum Qualitative Sozialforschung/Forum: Qualitative Social Research*, 6 (2), Art. 43. http://nbn-resolving.de/urn:nbn:de:0114-fqs0502430.

Minosi, V. 2005. Le eredità di Giovanni Urbani. *Arkos* 10: 26–30.

Moioli, R. 2004. Dalla pratica del cantiere alla teoria della conservazione. In A. Marchesi (ed.), *Villa Sottocasa in Vimercate*: 32–43. Missaglia: Edizioni Bellavite.

Moioli, R. 2009. La Conservazione programmata ed il progetto di restauro. In *Conservation Préventive. Pratique dans le domaine du patrimoine bati*, Actes du colloque, Fribourg, 3-4 septembre 2009: 161–167. Bern: SRC/SKR.

Moioli, R. 2015. Preventive and Planned Conservation and Economies of Scale. Conservation Process for 12 Churches. In K. Van Balen & A. Vandesande (eds) 2015. *Community involvement in heritage*: 103–116. Antwerp-Appeldoorn: Garant.

Moioli, R. & Baldioli, A. 2018. *Dieci Anni di Conservazione Programmata*, Quaderni dell'Osservatorio di Fondazione Cariplo, 29.

Nuccio, M. & Ponzini, D. 2017. What does a cultural district actually do? Critically reappraising 15 years of cultural district policy in Italy, *European Urban and Regional Studies* 24/4: 405–424.

Ponzini, D., Gugu, S. & Oppio, A. 2014. Is the concept of the cultural district appropriate for both analysis and policymaking? Two cases in Northern Italy. *City, Culture and Society* 5(2): 75–85.

Staniforth, S. (ed) 2013. Historical Perspectives on Preventive Conservation. Los Angeles: Getty Conservation Institute.

Therond, D. & Trigona, A. (eds) 2009. *Heritage and Beyond*. Strasbourg: Council of Europe Publishing.

Urbani, G., (2000). *Intorno al restauro*. Milano: Skira.

Van Balen, K. 2015. Preventive Conservation of Historic Buildings. *International Journal for Restoration of Buildings and Monuments*, 21 (2-3): 99–104.

Van Balen, K. 2017. Challenges that Preventive Conservation poses to the Cultural Heritage documentation field. *The International Archives of the Photogrammetry, Remote Sensing and Spatial Information Sciences, Volume XLII-2/W5, 2017 26th International CIPA Symposium 2017, 28 August–01 September 2017, Ottawa, Canada*: 713–717.

Van Balen K. & Vandesande A. (eds) 2013. *Reflections on Preventive Conservation, Maintenance and Monitoring of Monuments and Sites*. Leuven: Acco.

Van Balen K. & Vandesande A. (eds) 2016. *Heritage counts*. Antwerp-Appeldoorn: Garant.

Van Balen, K. & Vandesande, A. (eds) 2019. *Professionalism in the Built Heritage Sector*. Leiden: CRC Press/ Balkema.

Vandesande, A. 2017. *Preventive Conservation Strategy for Built Heritage Aimed at Sustainable Management and Local Development*. PhD. Dissertation, KU Leuven.

Van Roy, N. 2019. *Quality improvement of repair interventions on built heritage. A framework for quality improvement based on stakeholder collaboration through knowledge enhancement and continuous care*. PhD. Dissertation, KU Leuven.

Monumentenwacht model and new initiatives

Preventive conservation model applied in Slovakia to monitor built heritage damage

Pavol Ižvolt
Department of preventive maintenance, The Monuments Board of the Slovak Republic

ABSTRACT: The original ambition of the Ministry of culture of the Slovak Republic and Monuments Board of the Slovak Republic (Pamiatkový úrad SR) as its executive body responsible for the protection of the monuments in Slovakia, was to develop and implement a preventive maintenance model for immovable cultural heritage based on a Dutch system called Monumentenwacht. At its initial phase the system was developed by the project called "Pro Monumenta – prevention by maintenance". The article explains conservation state of the build protected monuments in Slovakia, organizational scheme of the used model, the most frequently occurring defects on the build heritage, struggled problems, development of the model in the years 2014-2019 along with the ideas and perspectives for the upcoming years.

1 INTRODUCTION

1.1 State of built heritage in Slovakia

Specific situations concerning monument protection have occurred mainly in countries with dramatic changes in ownership rights and social composition of the society. Relocations of inhabitants after World War II, nationalization of private establishments, neglected historical town centres and other factors made significant impacts on the status of monuments in Slovakia. New owners often had weak relation to historical sites, not even customs vital for their maintenance. Absence of preventive care of many historic sites led to theirs completely critical construction and technical conditions. When these monuments were in some cases finally renovated, they were frequently radically modernised and most of their monumental values connected with original fabric were lost as a result of replacing many original elements. Another contravening feature common also for other modern countries is a substitution of traditional crafts and manual work with industrial civil engineering and unified construction procedures and construction materials.

Currently, country registers 10 038 national cultural monuments (18 667 objects). Their constructional state is in principle not improving, but only maintained at approximately the same level, with hundreds of them decaying, not being used, or lacking basic maintenance.

According to the currently prepared interim report of the Revision of Cultural Expenditure, 25% of the monuments are in disrupted or desolate state and so-called "monument debt", that is the investment that would theoretically be necessary for its recovery. It amounts to 1.8 to 5.5 billion euros.

1.2 Core idea

It can be presumed that the idea of material monument protection is based on the societal decision to extend the life of its selected works (arts, architecture, milestones of technical development etc.), which creates both its heritage and environment. The built monuments are mostly exposed to weather, everyday use, natural disasters and changes in functionality, periods without use and periods of neglect. Damage to buildings does not apply their potential hazards and abundance equally to all types structures and not all materials used are equally durable. Emphasis on preventive control and maintenance of high-risk parts of the buildings, timely coating and proper ventilation can keep building structures in good construction condition admirably long.

The core idea of the Slovak project "Pro Monumenta – prevention by maintenance" and its ongoing model represents a simple concept expressed by Dutch Monumentenwacht demonstrating that regular and systematic maintenance may postpone or completely rule out financially–comprehensive conservation projects. It is at the same type the way to keep as much as possible of the original historic substance.

1.3 New challenges

Since the beginning of the Dutch Monumenten-wacht, the context for the preservation of monuments has slightly changed and new challenges have emerged.

Is maintenance efficient? The basic idea of maintenance as a precautionary measure is the verified knowledge that: regular maintenance can delay or even eliminate costly projects complex restoration of building monuments. Another justification for prioritizing maintenance is an ecological perspective. Investing in prevention and repair has a lower "carbon footprint" and lower environmental burden as a replacement of a building component or even replacement of the whole building by new one. We can continue with other challenges as global climate changes, natural disaster risks, changes in labor market etc. All these challenges can be answered by the modifications of the "classic" Monumentenwacht model. Maintenance projects can be developed in different variations and also collaborate with completely different departments (as social affairs, education, environmental projects etc.).

2 PROJECT "PRO MONUMENTA – PREVENTION BY MAINTENANCE"

2.1 First steps

Spreading the idea of monument maintenance abroad, the Dutch Monumentenwacht presented its inspection in Slovakia for the first time in 1999. The idea of preventive maintenance resonated since that time among Slovak conservators as a possible important element in the system of monument protection and one of the answers to the poor technical state of monuments.

In 2011, the Cultural Heritage Section of the Ministry of Culture of the Slovak Republic provided the Norwegian side with suggestions for a possible bilateral (pre-defined) project for the new planned European Economic Area (EEA) grant period. The Norwegian side preferred a common preventive maintenance project, because Norway did not have such a project to a greater extent and the Slovak experience might also be inspiring for Norway.

From the beginning, the Slovak side assumed that the most appropriate solution, unlike the Dutch model, would be to build a unit of preventive protection at the Monuments Board of the Slovak Republic - which is a state responsible office for the monument protection operating in the whole country.

2.2 Established model of the preventive maintenance

The Department of Preventive Conservation was established in January 2014 at the Monuments Board of the Slovak Republic. Altogether, it had 11 job

positions: project administrator, expert coordinator and 9 inspectors in the three independent working groups, evenly distributed geographically. They operated in the locations of regional monument boards in Trnava, Banská Štiavnica and Poprad-Spišská Sobota. These were and still are the venues of the teams' offices, their storehouses for auxiliary tools and material, and parking lots.

The first three years of the department's functioning were financially covered by EEA grants as a project called "Pro Monumenta – prevention by maintenance" (with allocated funds totaling euro 1,152,056). Since 2017 the work of the department has been financed by the Slovak Ministry of Culture (as a maintenance period). The new project with the aim to extend existing activities called "Pro Monumenta II." has started on the 2nd of May 2019 and should continue up to 30th of April 2022 (with allocated funds totaling euro 1 500 000. Then it will continue in its sustainability period till 2027.

Figure 1. Full equipped team of inspectors with theirs "mobile workshops".

2.3 Methodology of the workflow

The three-member team (since 2019 extended to four-member team) represents an independent unit in organising their work time, estimating the share of time spent on the site, time allocated for preparation and data processing and administration. Currently, the schedule can be roughly outlined as follows: time spent at the monument site 20%; project administration 20–30%; compiling technical reports, sorting photos, producing engineering drawings 50–60%. Leaving out days off, incapacity to work, training days and similar events, time devoted to one monument would cover one week. Members of the teams are employees of the Monuments Board of the Slovak Republic, forming part of The Department of Preventive Conservation. The centre of the Monuments Board of the Slovak Republic is in charge of sending the introductory letter and the final technical report to owners. The specialised project coordinator explores background documentation in archives, audits individual technical reports, elaborates methodological guides and takes part in producing lists of monuments for inspection.

The daily routine of an inspection group would comprise collection and studying of documentation pertaining to a particular monument, filing of documentation, scanning of engineering drawings, contacting the property owner, physical inspection and description of the site, cleaning of roof gutters (obligatory for the each inspection) and demonstration for the owner of handcraft procedures needed to repair minor defects. Vehicles utilised throughout the project implementation are called "mobile workshops" and are fully equipped with ladders, tables, tools and diagnostic apparatus and equipment for taking samples such as precise thermometers and hygrometers, thermal cameras, an endoscope, "Presler drill" (for dendrochronological samples) and so forth. The owner receives after its elaboration a technical report on the status of the cultural monument.

Roof accessibility plays an important role, thus the team is equipped for work at heights and performs the tasks while observing specific safety rules. In reality, it is often difficult to access all parts of the roof without scaffolding. Therefore one member of each team was specially trained for the work with drone. The use of drones shortens the time of the first gutter inspection and inspection of the roof defects. The upcoming physical intervention (cleaning and repair) is more accurate thanks to that.

While on inspection, staff members take pictures and use an original, tailor-made software application via tablets. This special application registers construction defects, lists photographs with indication marks, and allows the staff to retrieve any part of the manual, if needed, along with web pages. Moreover, it helps to make mainly visual recordings (photos, video), captures data on measurements performed by technical devices (thermal camera, endoscope, hygroscope). At the same time, the staff also prepares and adapts material for simple demonstration of repairs on problematic spots. If needed (for a distinguished monument or some part of it), samples can be taken for further analysis to resolve recurrent issues. Besides verbal description of defects, we also apply a scale of defect based on the one used so far by the Monuments Board of the Slovak Republic, which is an amended scale pursuant to the European standard for diagnostics. Rating of the monument technical status:

1. Good
2. Acceptable
3. Disturbed
4. Poor
5. Under renovation
6. Not assessed

The fifth category "under renovation" and the sixth category "not assessed" were added. Technical reports from cultural monument inspections contain an assessment of relevant repairs and interventions from the past, a technical description, a summary of all identified results. They contain many illustrative photographs, sometimes sketches and drawings.

Figure 3. Repair of the member of the wooden truss.

2.4 Team creation

Members of inspection teams are required to be physically fit, work at heights, be manually skilful and have acquired specialised knowledge in engineering technologies, construction details and materials. Each team consists of three members. Team work is divided; however, individual members have to be replaceable if needed. To enhance their knowledge

Figure 2. Trained inspector during the repair work.

and skills, staff members undergo extensive initial training and workshops lasting for several months. New staff members come from various backgrounds but most of them have already worked with monuments in terms of their diagnostics and handcrafts. Physical fitness and a driving licence are required as well. During the first months, the teams went through theoretical and practical training sessions to harmonise and complement their abilities and expertise.

Figure 4. Cleaning of the gutters as a common part of the inspection.

As a result, they are now able to work at heights, mend plumbing, replace roof tiles, perform basic masonry work, repair damaged wood rafters, and so forth while their roles in the process are interchangeable. Nonetheless, in terms of their job descriptions, one of them is an administrator focusing mainly on research documentation and plans, communicating with the owner and being chiefly responsible for completion of the technical report, reporting on travels, attendance at work, etc. Another member is in charge of sound status of tools and preparation of necessary materials and stock-taking. Finally, the third member takes care of the vehicle serving as the mobile workshop. The three-member team represents an independent unit in organising their work time, estimating the share of time spent on the site, time allocated for preparation and data processing and administration.

The team is well equipped with all the basic tools for diagnostics and small repairs. The inspectors found useful some other devices as infra surface thermometer, hygrometers, thermal camera and endoscope.

Figure 5. Supporting construction in Komjatice.

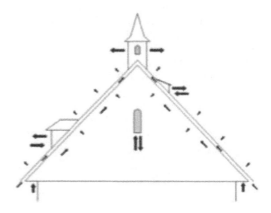

Figure 6. One of the illustrative schemes provided for the owners in the technical reports. The ventilation inside of the church building.

2.5 Selection of the monuments

In the attempt to make sure that as many as individuals and groups benefit from the project outputs – since it is funded from EEA Grants and state budget of the Slovak Republic – in its first implementation phase lasting until April 2017, the project was solely applied to cultural monuments owned by the state, local governments and churches. Simultaneously, the key was to select immovable cultural monuments that are, at least upon certain conditions, open to the public. Selection of sites was first made by the Monuments Board of the Slovak Republic, or respective regional monument boards, based on current information or the need for more information on the technical status of monuments within their territory.

Selection was based on evaluation of the "need" for such monitoring by the state, using several criteria such as historical value of the site, the period since its last renovation, capacity of its owner/administrator to remove potential defects, the need for expert evaluation or the necessity to acquire more precise information on the site, etc. With the aim of covering the broadest possible spectrum of monuments, the selection process took into consideration the type and location of the monument. Moreover, the selection made sure that the effectiveness and range was as wide as possible. Despite our efforts to diligently seek a diverse site typology, practice showed that the largest interest in inspection was registered among religious monuments, probably due to their dominant representation in the Central List of Monuments in Slovakia.

2.6 Most frequent defects

Despite the fact that long-term attention has been paid to construction failures in the literature, the daily practice of diagnosis brings interesting findings. In 4 years of field work, a large amount of practical observations and knowledge has been accumulated and an extensive database of failures has been created. The failures of historical buildings observed by our inspectors are in most cases related to the immediate surroundings of the buildings and to the water. Static structural failures are also largely due to changes in rainwater drainage, washing of foundations, especially through various drainage, gravel pits, water meter pits, concentrated soil infiltration under rainwater gutters and the like. Much of the remediation against rising damp does not work. The individual "fashion waves" of the solution usually brought more harm than good, for example the external ventilation ducts were flooded with rain water, the chimney effect of ventilation was not sufficiently dimensioned suction and exhaust, the cavity cooled the masonry in the interior. modifications of deep gravel beds with drainage gravel backfilling of foundations practically allowed water to soak up to the foundations quickly and to wash them down, which can lead to masonry collapse. The concreting of the eaves pavements increased the effect of the masonry capillary effect and also caused the spraying of skirting plasters with reflected rain drops. The latest fashion wave of widespread use of studded films does not improve the situation in any way. The cavity spaces of the foil during masonry are not sufficient to effectively ventilate any moisture, on the contrary, in practice, the foils usually do not have solved the overlap of the top bar and leak the amount of water running down the facade. In addition, the films are less flexible compared to conventional waterproofing membranes and cannot be used to copy the uneven edge of the foundation masonry. The issue of moisture remediation requires

a thorough understanding of each specific case and special specialization. However, there is a whole series of measures that can be carried out by the owners of the monuments themselves, for example to ensure consistent gradation of the terrain from the construction; check for clogging of drains and rainwater drains.

Figure 7. The "nightshift" – quick provisional protection of the roof in Vychylovka open air museum.

2.7 Collaboration with the Ministry of Culture

For the Ministry of Culture of the Slovak Republic the inspectors prepare evaluation reports on the state of cultural monuments, whose owners apply for a financial subsidy under the program Let's Renew Our House, Subprogram 1.6. For supported monuments inspectors regularly monitor the quality of construction work, on the basis of which the Ministry of Culture releases other subsidies. This work is useful for both parties. For the "Pro Monumenta team", this is a new experience with ongoing construction works and studying of architectural projects and on the other hand, the quality of repair works is improving.

2.8 New project "Pro Monumenta II"

At present (since May 2, 2019), a new project, Pro Monumenta II, is launched, bringing several innovations. Its main goal is to complete and existing system of preventive maintenance in a sustainable way. The duration of the project has been assumed to be 3 years.

Inclination for more intensive use and promotion of traditional building trades was reflected in the preparation of three craft training centers (workshops and conference centers) situated in Trnava, Levoča and Banská Štiavnica. These centers will provide room for preventive conservation teams, garages for theirs vehicles, stores for material and different training and lecture programs especially for administrators and owners of the monuments.

Another planned new service is the provision of preliminary economic services calculation of estimated repair costs. In the near future is also expected the use of drones, which can substantially save the time needed to anchor and secure the ropes for inspectors checking hard-to-reach parts of roofs. Cameras mounted on drones they can create orthogonal photographs of roofs, which are subsequently marked by fault points. New video tutorials for maintenance of selected parts of the monument will also be distributed through this web site.

As for the publicity maintenance a new event for the municipalities will be introduced - "Clean the gutter day". At least 4 local events will be organized in most important protected urban monument areas: Banská Štiavnica, Kežmarok, Levoča, Skalica with a close collaboration with local governments and local NGOs. All inspectors team will take a part in these promotion days organising free cleaning of gutters, inspections of roof constructions, consultations and recommendations. The campaign will promoted via local radio, TV and other media.

Figure 8. Inspection of the Oslo Cathedral by Pro Monumenta Slovak team.

The introduction of a system of preventive inspections of immovable cultural monuments in the Kingdom of Norway with elements of the Slovak system has been planned since the beginning. Two important churches – wooden church in Drobak and The Oslo Cathedral were inspected with the presence of Norwegian media, The Ministry of Environment and public.

Lithuania was also interested in the application of the Slovak system with its FIXUS project, which has been already started with Slovak collaboration.

3 CONCLUSION

3.1 Results and conclusions

The centralized model (based on the Monuments Board as a state organization) consisting of regional centers applied in Slovakia has been proved as suitable and well balanced up till 2020, more than 450 monuments were inspected, organizing of many seminars, public events, distributing several handbooks and publishing a guide for maintenance. Repair works were carried out during 237 inspection trips.

The most effective key factor which ensure the high quality of results is the well trained stuff.

The visual inspections are the main way of building damage assessment; however it should be completed with all possible up-to-date technologies, considering mainly their practical accessibility. Preventive maintenance systems can be in the present society extended or upgraded with new functions (support of traditional historic craft techniques, economic calculations of repairs, collaboration with the grants systems etc.).

3.2 Problems occurred during the implementation

Typical features of project financing require developing budget items well in advance, which results in little flexibility to respond to external events. Long-term funding restraints in the sphere of culture make some compensation for newly-arising requirements impossible. Public procurement of individual items (e. g. purchase of tailor-made "mobile workshops") is notoriously infamous for its lengthy procedure that hinders the logical sequence of other project components. Despite initial worries, many doubts concerning lack of interest on the part of owners or feasibility of training qualified personnel did not materialise. The topic of monument maintenance and diagnostics has not been sufficiently tackled from a theoretical perspective in our environment and detailed technical information had to be collected from various partial studies or complemented from abroad. The long-term viability of this new component in Slovakia's monument protection system will require some financial investments. A way forward might be multi-source financing, sponsoring, and involvement of monument owners by paying a fee for inspection, participation in research projects and close cooperation with local governments.

3.3 Major differences from the "classic model of Monumentenwacht" and characteristics of the Slovak model

In spite of the classic Monumentenwacht, the practice in Slovakia proved that the monuments are generally in more critical technical situation (more similar to the situation in the Netherlands before establishing of the Monumentenwacht in 60-ties of the 20th century). Therefore the time needed for first inspection including writing of the report is in Slovakia longer (5-10 days). The most of the inspectors are university educated with the practice in conservation and the manual craft too. Although the inspection teams have a wide variety of the electronic tools and modern equipment, in the time pressure they sometimes prefer to use simpler and more practical tools and working methods (visual assessments etc). In Netherlands and Flanders, the whole system operates on the level of regions, in Slovakia a central model has more advantages.

3.4 Long term goals of the preventive conservation model in Slovakia

Among the most anticipated goals that are ambitious but partially achievable are:

- to become a standard tool of heritage police widely popular in the society
- to keep the highest professional credit among the heritage professionals

- to improve the quality of restoration of monuments in Slovakia
- to rehabilitate traditional building craft
- to set up and collaboration with other possible heritage management projects (revolving fonds etc.).

REFERENCES

Ižvolt, P. 2017. Údržba historických stavieb. Príručka pre preventívnu údržbu nehnuteľných pamiatok – skúsenosti z projektu Pro Monumenta: 238. Bratislava: Pamiatkový úrad SR.

Ižvolt, P. 2015. Údržba alebo rekonštrukcia? Viac o projekte Pro Monumenta In Urbanita, roč. 27: 66–69. Bratislava.

Ižvolt, P. 2019. Preventive conservation model applied in Slovakia and the built heritage damage monitoring. In Preventive conservatie van klimaat – en schademonitoring naar een geïntegreerde systeembenadering. Leuven: WTA Nederland.

Ižvolt, P. 2014. Pravidelná údržba a jej filozofia. In Gembešová, L. – Ižvolt, P. – Kvasnicová, M. – Škrovina, M. – Urlandová, A. Škola remesiel. Tradičné stavebné remeslá – obnova historických drevených brán: 16–19. Svätý Jur: Academia Istropolitana Nova.

Conservation of cultural property – Condition survey of immovable heritage, CEN/TC 346, TC 346 WI 346013. Accessed January 2020 at http://euchic.eu/images/uploads/N042_WI_346013_(E)_Immovable%20cultural_2010-01-12_for_CEN_enquiry.pdf.

Inštitút kultúrnej politiky, 2019, priebežná správa. Revízia výdavkov na kultúru. Accessed January 2020 at http://www.culture.gov.sk/ministerstvo/institut-kulturnej-politiky-/revizia-vydavkov-33c.html.

*Preventive Conservation - From Climate and Damage Monitoring to a Systemic
and Integrated Approach – Vandesande, Verstrynge & Van Balen (eds)
© 2020 Taylor & Francis Group, London, ISBN 978-0-367-43548-6*

The Traditional Buildings Health Check: A new approach to the built heritage in Scotland

Sonya Linskaill

Trust Manager, Stirling City Heritage Trust, Stirling, Scotland, UK

ABSTRACT: The 2010 Scottish House Condition Survey, a national survey of housing and house-holds, found that 76% of traditional domestic buildings required repairs to Critical Elements (Scottish Government, 2011). However there was no national strategy to tackle poor maintenance and disrepair of Scotland's built heritage. In 2013, Historic Environment Scotland and the Construction Industry Training Board (Scotland) in partnership funded a 5-year pilot project to practically address this issue, and challenge cultural inactivity in this area. The Traditional Building Health Check is based on the European model for preventive maintenance, Monumentenwacht. Through building inspection and an educational ethos, the Traditional Building Health Check service aims to address the multiple and complex reasons for disrepair to Scotland's built heritage. The service supports property owners to proactively repair and maintain their buildings. It aims to challenge the acceptance of low quality, poorly executed repairs which result in deterioration of the building in particular and the cultural heritage in general.

1 INTRODUCTION

The findings of the Scottish House Condition Survey were of concern to a number Scottish agencies involved in the care and repair of Scotland's built heritage. Traditional buildings, as defined by the Survey, are those which predate 1919 and represent 20% of Scotland's housing stock, approximately half a million dwellings (Scottish Government, 2011). National statistics on commercial and public buildings are not collected. The Critical Elements are those parts of the building fabric, the condition of which is critical to a dwelling being wind and weather proof and structurally stable.

It is recognised that disrepair of the external building fabric has implications for Scotland's heritage, both culturally and materially. Such implications include:

- Performance, e.g. in extreme weather.
- Wellbeing of the occupants, e.g. air quality.
- Public safety e.g. masonry collapse.
- Energy efficiency and fuel poverty.
- Attractiveness: physical, social and economic.
- Scotland's sense of place and identity.

The causes of disrepair extend from a lack of planned preventive maintenance and timely repair, to inappropriate and poorly executed repairs and design interventions.

2 DESCRIPTION OF THE PILOT PROJECT

2.1 *A Monumentenwacht for Scotland?*

In 2013, two agencies with an interest in the protection and repair of Scotland's built heritage formed a partnership, and committed funding to a pilot project to practically address the repair and maintenance of the country's traditional building stock. These agencies were Historic Environment Scotland, a non-departmental public body and the lead public body for the country's historic environment, and the Construction Industry Training Board (Scotland). The latter works to ensure that construction employers have the right skills, in the right place, at the right time, by investing funds and providing a wide range of industry-led skills and training solutions.

The resulting 5-year pilot project, the Traditional Buildings Health Check, was based on the European Monumentenwacht model in the Netherlands and Flanders, and adapted to fit within the context of the Scottish heritage environment and the challenges and objectives therein. One significant difference between the original European model and the Scottish one, is that the Scottish pilot was limited to owners of traditional buildings in the selected pilot area of Stirling and not eligible to the rest of Scotland.

The pilot was delivered by Stirling City Heritage Trust, a local heritage agency and registered Scottish Charity. Launched in December 2004, the Trust was set up by Historic Environment Scotland in partnership with the local authority (Stirling Council) as a means of delivering focused investment in the historic built environment of the City of Stirling.

2.2 *Stirling*

Stirling, a small city with a population of approximately 36,000 (Scottish Government, 2011), was selected for a number of reasons. An historic Royal burgh and major royal residence from the 14th to 16th centuries, Stirling contains eleven conservation areas and almost five hundred listed buildings. It comprises a diverse mix of property types in its urban centre and outlying residential areas, with a mixed demographic. For example, Stirling Town & Royal Park Conservation Area was within the 15% most deprived areas in Scotland (Stirling Council, 2012). This is the area immediately adjacent to the internationally important Stirling Castle and contains the majority of the city's assets of national importance. It could provide a suitable example of the diversity of Scotland's historic built environment.

2.3 *Traditional Buildings Health Check*

The pilot project established a membership based service providing impartial and expert advice on the maintenance and repair of the external fabric of traditionally constructed buildings. For a modest annual member subscription fee, building owners had access to a subsidised inspection service carried out by a team of two traditional building inspectors. A detailed and illustrated report on the property's condition was provided with prioritised recommendations for repair. Meetings were offered to owners to discuss the results of their inspection and assist in their decision making on appropriate and timely repair of the building fabric.

The inspection report was recognised as the principal mechanism to support understanding of and action on the building fabric. Defects were prioritised using three categories:

Priority 1: Repairs, replacement or recommended investigations should be carried out at the first available opportunity and in less than 12 months. Failure to attend to these defects may cause further deterioration or damage elsewhere or create a safety hazard.

Priority 2: Repairs, replacement or recommended investigations are required and works can be phased to make best use of resources and any access required. If not attended to these defects may be elevated to a higher priority.

Priority 3: The condition of the building element is consistent with age. No repair or replacement is currently needed. The element should be regularly inspected and maintained as necessary to reduce the possibility of failure.

The simplification of the priorities into three categories was to enable property owners to focus on the most crucial aspects which required urgent attention, whilst being aware of other potential repairs in the short to medium term. The categorisation deviated from BS7913:2013 (BSI Standards Limited, 2013), in a practical decision by the project team to create a system which was as straightforward as possible for an owner without any knowledge of traditional building methods and materials. The categories were graphically represented by the red-amber-green protocol of road traffic signals.

The inspection and reporting service met the pilot project's first short-term objective: to provide a condition assessment of the building and recommendations on repair and maintenance. The second objective was to stop decay by carrying out preventive maintenance and urgent repair. To achieve this second objective, the Traditional Building Health Check team required to engage further with individual property owners using the information gathered during the inspection process. There were six key steps:

1. Identify defects and analyse the causes of decay and disrepair.
2. Owners take responsibility and acknowledge that their building is in a state of disrepair, and furthermore understanding what is required.
3. Prioritise and schedule interventions.
4. Design specific appropriate interventions as required to ensure sustainable details, using suitable materials and methods for the intervention.
5. Successfully intervene using competent contractors with suitable skills and expertise.
6. Continue preventive maintenance and consider intervention through regular re-inspection and planned works.

Figure 1. TBHC Inspector on site.

3 FINDINGS OF THE PILOT PROJECT

3.1 Levels of disrepair

Almost three hundred members joined the Traditional Buildings Health Check during the pilot from its operational launch in October 2014 to March 2018. 255 of them had a building inspection carried out, providing a significant base of physical evidence.

Data collected during the inspection process allowed comparison between 'Priority 1 defects' identified in Traditional Buildings Health Check reports, with 'Disrepair to Critical Elements' identified in the Scottish House Condition Survey (Scottish Government, 2011). The pilot data confirmed the national statistics of the Survey, as virtually all buildings had some elements in need of repair, with 88% requiring work in the next 12 months (Stirling City Heritage Trust, 2017).

3.2 Causes of disrepair

The Traditional Buildings Health Check found that the causes of disrepair could be divided into two types. Firstly those which were the result of natural deterioration of the original building fabric and a lack of timely repair. A particularly common example being the long-term neglect of high level elements such as chimneys (Figure 2). Secondly inspectors found significant levels of accelerated and hidden disrepair resulting from poor quality or inappropriate interventions (Figure 3). This included repairs using inappropriate materials; temporary repairs left over an extended period; poorly executed repairs demonstrating a lack of knowledge and skills on traditional building methods; and poorly designed interventions to address inherent problems of some original details (Figure 4). In many cases the poor interventions were a combination of more than one of these issues.

In addition to the physical evidence of disrepair, working closely with members identified other causes. For instance a lack of knowledge amongst

Figure 3. Typical poor quality repair of a natural sandstone chimney with an inappropriate cement based render instead of suitable stone repair (stone indent and replacement).

Figure 4. Poorly designed intervention using lead sheet to cover original stone copes to solve perceived water penetration through the copes.

the owners, and insufficient numbers of skilled contractors, compounded inactivity and low quality, poorly executed repairs.

4 DISCUSSIONS

4.1 Disrepair

The Traditional Buildings Health Check pilot, using the typical historic built environment of the City of Stirling, corroborated the level of disrepair of Scotland's traditional buildings. Although the original slogan of the service was to 'maintain your building', the pilot found out that its main focus was to encourage proactive appropriate repair.

4.2 Poor quality interventions

The extent of poor quality interventions was larger than anticipated. Many contractors working on

Figure 2. An example of long term neglect of a chimney.

traditional buildings would appear to be inadequately skilled in the use of appropriate materials and working methods, in particular traditional crafts such as lead work and the use of lime mortars. Probably as expected, owners often had little knowledge of traditional building methods and materials and therefore did not know what to ask from the contractors, or how to assess the completed works. On small repair works there was often no professional involvement from an architect or surveyor, and no legal requirement for permissions which would have allowed the local authority to assess the proposed works.

A survey of members undertaken during the pilot indicated that a third (35%) of those surveyed said that the Traditional Buildings Health Check report supported them to carry out the appropriate interventions in due time; a further third (29%) had done additional work as a result of the report; 39% of members stated that they were aware of problems, and recognised they needed professional advice (Stirling City Heritage Trust, 2017).

4.3 Cultural attitudes

An estimated £1.3 million of private investment was made by Traditional Buildings Health Check members for the repair and maintenance of their properties over the pilot period (Jura Consultants, 2018), a tangible benefit to the traditional building stock of Stirling and the local economy. This investment suggests that the willingness to repair is higher than first assumed and that other factors such as education and availability of competent contractors, and in some cases building professionals, were more likely contributory factors for inactivity.

The service tackled inactivity caused by the lack of knowledge, by providing expert advice and identifying the most suitable intervention. The survey found that 44% of Traditional Buildings Health Check members subscribed to find out if there were problems with their property which they were unaware of. At least 55% of members had maintained or repaired their buildings since subscribing to the service (Stirling City Heritage Trust, 2017). The survey therefore indicated that the Traditional Buildings Health Check could act as a catalyst for appropriate repair and maintenance of traditional buildings.

5 CONCLUSION

The medium and long-term objectives at the beginning of the pilot included firstly to stimulate and stabilise the traditional repair and maintenance sector by gradually changing owners' attitudes to preventive maintenance. Secondly, the pilot aimed to provide owners with an adequate understanding of traditional building fabric so they would have confidence to undertake and commission repair and maintenance. The pilot project demonstrated that there is a valid and verified role for the Traditional Buildings Health Check service to provide impartial expert advice to building owners to ensure that they can make timely and informed decisions.

The intention remains to stimulate demand for a construction sector with the skills and capacity to appropriately maintain and repair Scotland's built heritage. This would ensure that the practical and technical resources required to protect the country's cultural built heritage exist, and that private and public investment in the proactive repair and preventive maintenance of traditional buildings is made in a sustainable manner.

Figure 5. TBHC Inspector on site assessing chimney repairs, a service offered to members throughout the year.

REFERENCES

British Standards Institute. 2013. *BS7913:2013, Guide to the conservation of historic buildings*. BSI Standards Limited.

Jura Consultants. 2018. *Traditional Building Health Check Review, Final Report*. Unpublished.

Stirling City Heritage Trust. 2017. *Traditional Buildings Health Check, Interim Review to Historic Environment Scotland*. Unpublished.

Stirling Council. 2012. *Briefing Note: Scottish Index of Multiple Deprivation 2012: Castle and St Mary's (Zone S01006124)*. Unpublished.

Scottish Government. 2011. *Scottish Housing Condition Survey Key Findings 2010*. Accessed July 2019 at https://www.webarchive.org.uk/wayback/archive/20170701210348/http://www.gov.scot/Publications/2011/11/23172215/0.

Scottish Government. 2010. *Census 2011*. Accessed July 2019 at https://www.scotlandscensus.gov.uk/ods-web/area.html.

Preventive Conservation - From Climate and Damage Monitoring to a Systemic and Integrated Approach – Vandesande, Verstrynge & Van Balen (eds)
© 2020 Taylor & Francis Group, London, ISBN 978-0-367-43548-6

Quality of restoration of monuments: The role of Monumentenwacht

Silvia Naldini
Department AE+T, Faculty of Architecture and the Built Environment, TU Delft, Delft, The Netherlands

Geert van de Varst
Monumentenwacht Limburg, The Netherlands

Sanne de Koning
Monumentenwacht Noord Brabant, The Netherlands

Ernst van de Grijp
Monumentenwacht Gelderland, The Netherlands

ABSTRACT: A program on Professionalism has been started in the Netherlands aiming at enhancing the quality of interventions in monumental buildings, (listed monuments and historic buildings). Branches involved, like contractors and architects, create their own guidelines under the guidance of the Dutch foundation for a recognized restoration standard in the preservation and restoration of historic buildings and sites. Guidelines for the condition assessment of monuments are partly based on those originally developed by Monumentenwacht. This organisation fulfils the task of advising both owners and provincial government. The inspectors also evaluate the work done by contractors, signalling poor executions. This paper analyses the contribution of Monumentenwacht to enhance the quality of the interventions. Case studies are used for discussing potentials and limits of their practice. The project 'Monumentenwacht moves', in co-operation with Delft University of Technology, aims at creating a uniformity in reporting and monitoring. This co-operation lies within the framework of synergistic activities in the field of monument conservation carried out by of various institutions.

1 INTRODUCTION

The quality of restoration in the Netherlands is a complex matter, involving different actors, and is the focus of various programmatic activities. Preventive conservation, good practices and professionalism play a fundamental role therein. Equally important is the formulation of clear policies and values and their translation to practice, which concerns both technical actors and laymen, i.e. owners of monumental buildings. The importance and economic benefit of preventive conservation of the built heritage has been stated in various studies (Vandesande 2017) and the necessity of quality of interventions addressed (Roy Van 2018). In the recently concluded JPI 'Changes' project, involving the Politecnico of Milan (Italy), the KU Leuven (Belgium), Delft University of Technology (the Netherlands), and Gothenburg University (Sweden), best practices in conservation were investigated, both in theory and in practice (Changes, 2017; Vandesande et al. 2018). In The Netherlands the Monumentenwacht organization (North Brabant province) was associated partner in the project and also object of the research, because it had played an important role in preventive conservation since its creation in the 1970's. The organization operates in contact with owners, local government, architects, contractors and specialists and conveys technical and ethical principles of conservation to the parties involved. The ethical principles are rooted in the philosophy of interventions starting from the Charter of Venice (The Venice Charter 1964) and imply that conservation of the existent should have the priority. Its activities, tasks and responsibilities have changed over time, as a reaction and an adjustment to policy changes and the decentralization process of the main Dutch body for the care of monuments, the Netherlands' Cultural Heritage Agency (RCE). To support Monumentenwacht fitting in the new profile, a co-operation program with Delft University of Technology has started, 'Monumentenwacht moves', which is part of a broader context of synergistic activities of RCE, TU Delft and TNO (Netherlands Organization for Applied Scientific Research) in the field of monument conservation. The potential of Monumentenwacht has become clearer through case studies (Heinemann & Naldini 2018), which have also highlighted the aspects where more structure could increase efficiency and quality, making the organization better comply with its new tasks.

The actual contribution to the quality of the interventions to monuments is explored in this paper as well as the potential for further development. The Monumentenwacht Inspection report (manual) was referred to for the development of guidelines for condition assessment of monuments within the national program on the Quality of Restauration and Professionalism. This program is stimulated by the RCE (in 2010) and its realization entrusted to ERM, the Dutch foundation for a recognized restoration standard in the preservation and restoration of historic buildings and sites (Naldini & Hunen 2019; Inspection guidelines URL 2005). Decisions on interventions are ranked in terms of impact on the monument and priority is given to conservation

2 MONUMENTENWACHT: THE ADVISORY ROLE

The role and work of Monumentenwacht North Brabant, Limburg and Gelderland are discussed on the basis of the results of case studies and the analysis of inspection reports. The basis is further laid for a synergistic action aiming a better contributing to the quality of conservation at a national level.

Active in The Netherlands since the 1970's in the field of preventive conservation at a provincial level, Monumentenwacht carries out condition assessments of monuments on a regular basis and supports owners and other actors, like architects, when interventions are needed. The organization is non-profit and independent. The inspectors are craftsmen trained in monument conservation. As such they can perform some minor interventions and repair some damage, avoiding unwanted consequences. Monumentenwacht has grown to become an important organization, spread all over the country and trusted by its members (owners of monumental buildings). Nowadays the organization has taken up some tasks and responsibilities which used to be fulfilled by the RCE and has been informally assigned an advisory role, especially by private owners. Giving advice, however, may imply the risk of involvement with the market. That the organization maintains an independent role is deemed necessary by the Provinces for developing policies based on its annual reports. Via subsidization, freeing the organization from the need of finding other financial support, this risk can be avoided. It is further important to consider that giving advice is a very complex job, implying skills and systematic methods, and that Monumentenwacht inspectors have grown into an advisory role over time operating on the basis of extensive experience centered on visual inspections. The study of the theory as expressed by Monumentenwacht's handbook and its practical application in case studies has shown some critical points. The way of operating of Monumentenwacht is practice-oriented and the activities are carried out following sound ethical principles. The main issues are to clearly establish the limits of the visual inspections and to indicate sound assessment methods.

2.1 Contributing to enhancing the quality of restoration

Monumentenwacht carries out visual inspections governed by the ethics of conservation, meaning that in deciding on the necessary intervention priority is given to the preservation of original material and techniques. Often this also means that only strictly necessary actions are undertaken, with a clear economical advantage. Some cases are briefly addressed to show how the organization contributes to professionalism in conservation.

2.1.1 *The advantages of a professional co-operation*
The case of a monumental country house in the Gelderland region shows the importance of a good co-operation among professionals (in this case inspectors and contractors) in restoration. This aspect is of great relevance for guaranteeing quality in conservation (Roy Van 2018).

Under guidance of a housing corporation a type of paint, meant to prevent water penetration, but not compatible with the support, was applied to the external wall surfaces. In a short time, the most exposed parts like pinnacles and pediments showed damage, which soon spread along the whole surface. Asked for advice, Monumentenwacht inspectors showed that small cracks and jointures allowed water penetration. Moisture gathered under the paint, which prevented evaporation. Damage consisted in blistering and peeling of the paint and sanding of the plaster (Figure 1). A decision on the intervention was taken by the inspectors together with the owner and a certified (specialized) contractor.

The paint on parts showing damage after 4 years' service was removed, the plaster was allowed

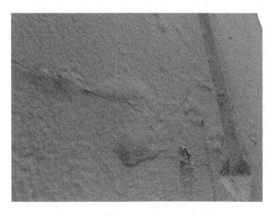

Figure 1. Blistering of paint (incompatible paint).

to dry and finally a silicate-based paint was applied. Following the advice of Monumentenwacht, lead plates were applied on the top of the most exposed parts, like the pinnacles, to prevent ingress of water. The intervention was reversible and gave good results. No damage has been observed so far (Figure 2).

2.1.2 Controls on quality

An interesting case concerning the repointing of a monument shows that a professional teamwork can provide the ideal approach in conservation.

The facades of a 19th century brick masonry monumental farmhouse also in the Gelderland region were (originally) furnished with a pointing showing a penny stroke and painted to match the shade of the brick (Figure 3). Due to change in taste, the most representative façade was repointed with a 'cut to shape' pointing, later in the 19th century (Figure 4). Reacting to the proposal of the contractor of cleaning the façade, Monumentenwacht has advised against the risks of this intended action, pleading the preservation of both the original pointing and the one renewed in the same century. This is a valuable example where the original colored and tooled pointing is still visible and can be preserved and even the cut to shape one indicates a change in taste in the past.

This is a valuable example where the original colored and tooled pointing is still visible and can be preserved and even the cut to shape one indicates

Figuur 3. Original (1804) pointing colored and furnished with penny stroke.

Figuur 4. Cut to shape pointing (end 19th century) (cf. Figure 3, same building).

a change in taste in the past. In many cases now a days the owner has the original pointing substituted with a modern cut to shape one, not considering the

Figure 2. The intervention (suitable paint and lead protective layer) proved successful (cf. Figure 1, same building).

Figuur 5. Original situation, pointing in good condition.

Figuur 6. Non necessary substitution, cut to shape pointing (cf. Figure 5, same building).

Figuur 7. Local activity of insects.

historical value (Naldini et al. 2001), as required by ERM guidelines (Figure 5, 6).

Following the *Kennis and Kunde* rules (Naldini & Hunen 2019), Monumentenwacht advises against hiring a contractor not considering the historic value of the buildings. The owner however carries the responsibility and finally makes a decision. In the case illustrated (Figure 6) also the cleaning of the façade permanently altered the original aspect of the masonry.

2.1.3 *Contrasting opinions between Monumentenwacht and contractor*

A problem often encountered by Monumentenwacht is the lack of supervision between the identification of damage and the suggestion on how to approach it and the interpretation of the contractor. Ethical principles founded in the need of preserving the past for future generation should guide the interventions. Instead choices are made towards not necessary and even damaging solutions.

Traces of insect activity were assessed in the original timber construction of the roof in a castle. Because of the local character of the damage, Monumentenwacht had advised to perform a very limited intervention (foot of the truss). Instead, the whole timber truss was treated, explaining the decision in terms of preventive conservation. The unnecessarily extended intervention included injections holes causing permanent alteration of the element (Figure 7, 8).

2.1.4 *Monumentenwacht assesses the work done by contractors*

For a long time Monumentenwacht had reported the need of intervention on the drainage system of a monumental building and had also performed minor repairs to avoid leakages before action was undertaken. After the intervention a new inspection was performed and the quality of the work was poor, because the technical principles of the connection

Figuur 8. Unnecessarily extended intervention (cf. Figure 7, same building).

Figuur 9. Principle not understood.

details were not well understood. As shown on Figure 9 the expansion element is incorrectly welded, and does not allow thermal expansion. The control of interventions is a relevant means of improving preventive conservation.

2.1.5 Reporting risk situations

According to the Dutch Heritage Legislation (Heritage legislation 2017) the owner of a building is responsible for its maintenance, that is to say that he needs to find the means to have it well preserved. The Municipalities exert controls and can induce reticent owners to comply with the requirements. Monumentenwacht is an organization supporting private persons and indicating priority of interventions, but does not operate for the Municipality. However, in the case that the condition state of (parts of) a monument can cause a safety risk to passers-by Monumentenwacht will strongly advice owners to take action.

Thanks to their regular inspections of roofs and higher parts of the buildings, Monumentenwacht could timely indicate severe damage to several cornices in the form of exfoliation and spalling (Figure 10). Being a safety-risk situation, action was promptly undertaken and the problems were consequently solved.

2.2 Monumentenwacht and building contractors

The presented examples of the work of Monumentenwacht presented show the contribution brought to quality in conservation. Very important part of Monumentenwacht's activities is to convince the owners of the need to pursue quality in intervention. This implies the choice of a contractor expert in restoration and controls of the plans of interventions based on the damage observed. Contractors play an important role in planning and carrying out interventions.

A survey carried out in 2018 by Monumentenwacht of the Gelderland province among their members showed that (building) contractors are considered by most of the respondents as advisors (Figure 11). Monumentenwacht plays an important role in enhancing the quality of conservation. However, in the work of the organization some improvements can be made in the way of reporting damage

Monumentenwacht Gelderland province – Survey question 11. Do you refer to other advisors than Monumentenwacht? Respondents: 171

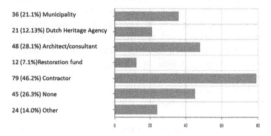

Figure 11. Survey Monumentenwacht Gelderland – 2018.

and making hypotheses on the damage cause. This applies for example to the use of a standardized damage terminology and a methodology for making hypotheses on which a correct diagnosis can be further based.

3 THEORY AND PRACTICE OF CONSERVATION: DIAGNOSIS

An ideal damage assessment includes visual inspection → hypotheses on causes of damage → control and diagnosis. In the practice of Monumentenwacht, maintenance advice is in clear cases based upon the visual inspection, without further investigations. In complex cases, on the basis of *visual inspections* Monumentenwacht can only formulate hypotheses on the damage cause(s). It is fundamental that Monumentenwacht explains to the owners that, in complex cases, specialist's advice and laboratory investigations may be needed to reach a sound diagnosis. In other words it is important to make a thorough visual inspection and be aware of its limits to come to a sound diagnosis.

3.1 Damage identification and causes of damage

Based on the evaluation of Monumentenwacht reports on inspected monuments the need of using a common and more specific terminology for the identification of damage has become clear. The clients for whom the reports are produced are usually laymen and this explains the use of expressions like *'some pointing weathered and locally partly washed out'*. However, such expressions do not identify the type of damage. Only the use of a uniform terminology for damage identification could guarantee good communication and further allow visual monitoring and control of the effectiveness of interventions. The program 'Monumentenwacht moves' has thus started with the use of a uniform terminology based on MDCS (Monument Diagnosis and Conservation System) (MDCS). This online system for damage assessment, diagnosis and interventions presents atlases including types of

Figure 10. Severe damage and safety-risk for passers-by.

damage and definitions, further related to possible causes. Within a following workshop, the theory on damage mechanisms supported by scientific knowledge will be applied to cases in practice to generate hypotheses: these will be based on the visual analysis of the materials used and their damage within given environmental circumstances. This method applies to more complex cases of damage, where not only ordinary maintenance is needed.

It is essential that inspectors correctly report on their inspections making clear what the limits of a visual assessment are.

When investigations (in the laboratory) are needed to control the hypotheses, the owner often relies on the contractor for carrying them out (Naldini & Hunen, 2019). Monumentenwacht should make it clear that further scientific specialized help and investigations may sometimes be necessary. In a further phase of the co-operation 'Monumentenwacht moves' technical support will be given by the TU Delft to Monumentenwacht to carry out investigations on moisture and salt. This action is meant to increase the level and the independency of the advice given.

Laymen owners trust and rely on the inspectors and this allows them to better involve the owners in the conservation process, encouraging them to seek professional contractors, to achieve sound interventions, yet preserving original materials. Owners should be encouraged to keep reports on inspections and ask contractors for reports on the intervention done to allow the evaluation of the choices made.

4 CONCLUSIONS

The cooperation of different parties involved in conservation is expected to be effective not only for the creation of technical guidelines to enhance quality, but also common values and principles to share. Being a widely spread and trusted organization Monumentenwacht not only provides technical assistance but also conveys these values and principles to other actors, like contractors and owners. In their activities the inspectors also contribute to enhancing the quality of interventions.

A very important role of Monumentenwacht is to make private owners aware of the value of their monuments - also at the level of the building materials and techniques - and of the importance and means to preserve it. This will contribute to the achievement of a sound plan of intervention to be carried out by contractors following specific technical guidelines for heritage conservation. The use of a clear, common terminology will create the basis for making well-structured hypotheses on the causes of the damage found, will improve communication and allow better visual monitoring.

The use of the structured way of operating developed in 'Monumentenwacht moves' is meant to allow the organization to better adapt to changes in conservation policy and where necessary to involve scientific research in their practical work.

REFERENCES

CHANGES 2015-2017, Changes in Cultural Heritage Activities: New Goals and Benefits for Economy and Society (2015-'17), http://www.changes-project.eu/ retrieved July 2019.

Heinemann H. A. & Naldini S., *The role and potentials of Monumentenwacht: 40 years theory and practice in the Netherlands*, in 'Innovative Built Heritage Models', eds. K. Van Balen & A. Vandesande, CRC Press/Balkema of the Taylor&Francis Group, Leiden 2018, 3: 107–115, special issue of the 'Journal of Cultural Heritage Management and Sustainable Development.

Heritage legislation (Erfgoedwet) 2017, https://wetten.over heid.nl/BWBR0037521/2017-09-01, retrieved July 2019.

Inspection guidelines URL 2005 https://www.stichtingerm. nl/richtlijnen/url2005], retrieved July 2019.

MDCS Https://mdcs.monumentenkennis.nl/ retrieved July 2019.

Naldini S., Hees R.P.J. van, Luxan M. Pilar de, Dorrego F., Balen K.E.P. Van, Hayen R., Binda L. and Baronio G., *Historical pointing and the preservation of its value*, in Proceeding of the Congress 'Structural Studies, Repairs and Maintenance of Historical Buildings VII', Bologna, 2001, WIT-Press Southampton, Boston 2001: 671–680.

Naldini S., Hunen van M., *Guidelines for quality of interventions in built Cultural Heritage* CRC Press/Balkema of the Taylor&Francis Group, 'Journal of Cultural Heritage Management and Sustainable Development', 2019:. 87–93.

Roy Van, N., *Quality improvement of repair interventions on built heritage*, PhD, KU Leuven 2018.

The Venice Charter 1964 https://www.icomos.org/charters/ venice_e.pdf, accessed July 2019.

Vanderzande, A. *Preventive Conservation Strategy*, PhD, KU Leuven, 2017.

Vandesande, A., Van Balen, K., Della Torre, S., Cardoso, F. *Preventive and planned conservation as a new management approach for built heritage: from a physical health check to empowering communities and activating (lost) traditions for local sustainable development*. Journal of Cultural Heritage Management and Sustainable Development, 2018, 8, 2: 78–81.

Preventive Conservation - From Climate and Damage Monitoring to a Systemic and Integrated Approach – Vandesande, Verstrynge & Van Balen (eds)
© 2020 Taylor & Francis Group, London, ISBN 978-0-367-43548-6

Preventive and planned conservation for built heritage. Applied research in the University of Porto

T.C. Ferreira
Centre for Studies in Architecture and Urbanism, Faculty of Architecture, University of Porto, Porto, Portugal

ABSTRACT: This paper presents research on preventive and planned conservation strategies applied to a specific case study in the University of Porto, the Faculty of Architecture, designed by Álvaro Siza and built between 1985 and 1993. The research methodology bridges the 'material component' (focused on buildings) with the 'intangible component' (focused on users), acknowledging that preventive and planned conservation are increasingly successful when linked with the participation and empowerment of users.

1 INTRODUCTION

"Maintenance (…) means not allowing a building to decay. Firstly, it's only about small tasks, but often we let things progress until a profound state of decay. Then, it is not about maintenance anymore but about rehabilitation, and rehabilitation is expensive. So, carrying out constant maintenance is also a question of economics. (…)." (Siza, 2017).

As Álvaro Siza maintains, the present-day context calls for a careful management of economic and ecological resources, shifting from a reactive and interventionist approach (post-damage) to preventive conservation (pre-damage) and continued care over time (Della Torre 2003, Van Ballen & Vandersande 2013, among others). Hence, this approach provides for a more sustainable management of resources, because of its cost effectiveness (decrease of costs over the medium and long term), reducing risk and damages, ensuring the preservation of the authenticity and integrity of buildings, fostering the participation of the users, and increasing their self-esteem in relation with built heritage (Ferreira 2014a).

This paper uses the broad concept of preventive and planned conservation (PPC) (Vandesande et. al), enclosing different kind of actions such as inspection, monitoring, cleaning, repair, use, etc.).

However, there are still very few practical implementations of PPC in Portugal (Ferreira 2018a), so the purpose of this paper is to present applied research to a specific case study in the University of Porto, the Faculty of Architecture designed by Álvaro Siza and built between 1985 and 1993.

Hence, this paper allows to demonstrate the benefits of PPC in complex buildings with intensive use and occupation such as Schools. Also, it explores the implementation of this strategies in late-modern buildings which, in certain instances and for different reasons, are generally more vulnerable than traditional ones because of its design and construction features (flat roofs, wide windows, experimental materials and solutions, etc.) (Canziani 2009; Ferreira, 2014b). Moreover, this research presents experiments of participatory strategies, acknowledges that only with the participation and empowerment of all users can we achieve successful and sustainable management of built heritage (Ferreira, 2018b).

2 METHODOLOGY

2.1 *Research methodology*

Following previous research (Ferreira 2014a, 2017, 2018), the methodology follows two complementary action lines: (1) 'material component' (physical intervention) focused on buildings – through inspection, monitoring and repair; (2) 'intangible component' (participatory strategies) focused on users – through education and training on prevention, maintenance and use.

The methodologic framework envisages different kind of actions focused on a comprehensive knowledge of the life cycle of the building: bibliographic and archive research (original project and chronology of interventions), systematized interviews to users (managers, staff, maintenance and cleaning staff, teachers, students), inspection and monitoring with NDT (termo-higrometer, thermography, decay mapping, visual inspection), field work and data collection (users, materials and techniques, anomalies, etc.), drawing of constructive details, consultation of contractors for maintenance planning, design of infographic illustrated manuals for users, among others.

All this data is compiled in a digital database and computerized software and App (MPLAN – Management & Maintenance Planning) specifically conceived for the PPC of built heritage.

2.2 Digital database and software (MPLAN)

MPLAN provides for different functionalities such as Management of Information & Database, Inspection & Diagnosis, Budget Estimation, Maintenance and User's Manuals, Schedule of Maintenance Tasks, Alerts & Notifications in real time, among others.

One of the main features of this digital tool is the compilation of Maintenance Plans, which are divided into five sections: Identification, Characterisation, Diagnosis, Maintenance, and Utilization. The Identification section (I) includes general information about the Building and the Users. The Characterisation section (II) contains the Characterisation Form and Building's Drawings, as well as a Chronology with building's previous interventions. The Diagnosis section (III) comprises the Anomalies Form, the Utilisation Form and the Log of Occurrences reported by users. This section also allows for the uploading of different kinds of NDT tests and analyses, among others: thermo-hygrometric or structural monitoring, decay mapping, thermographic captions or inspection reports. The Maintenance section (IV) relates to the scheduling and description of inspection, maintenance and repairs (Maintenance Manual) and Intervention Form. Finally, the Utilization section (V) provides for Users Manuals including recommendations for Use and Cleaning.

Many of these forms have pre-compiled or dropdown fields in order to facilitate their rapid completion and reduce errors. There is also a detailed Tutorial to help users filling in the different forms, providing for lists and descriptions (materials, constructive systems, damage atlas, repair actions, users' recommendations, etc.). The software has reserved access through different logins (Administration, Manager, Technician, User) with distinct permissions, including a system of Notifications & Alerts in real time. Besides the compilation of Maintenance Plans, MPLAN allows for the schedule of preventive and reactive tasks, annual budget estimation, advanced research fields. It also consents the possibility to link with other facility management systems or visualization tools (BIM, 3D models, virtual tours, etc.).

MPLAN is available for desktop and mobile devices in order to facilitate completion and consultation in field work. Digital tools such as MPLAN are essential for PPC as they allow for time and costs saving, providing for updated reports and database.

Figure 1. MPLAN Software Menu (the author).

3 APPLIED RESEARCH IN THE UNIVERSITY OF PORTO

The University of Porto (UP) is one of the largest education and research institutions in Portugal (31.530 students in 2016 and owns several buildings in different campus and areas of the city, as well as in the north region (150 properties, 511.191m^2 built, and 864.855m^2 of private land use).

The responsibility of building maintenance is related to each Faculty and linked with the Service of Installations and Infrastructures of the Rectorate of the UP. This Service has been providing support to the single Faculties, either in funding conservation works or in adapting installations and equipment to the new legal requirements and maintenance conditions. With this scope, annually (e.g. 2015), there are approximately 200 procedures for works in buildings of the UP, in a total amount of 22M, which have to respond to the constraints of the Portuguese law for public contracts (Ramos 2017).

3.1 Good practices illustrated manuals

The Rectorate has recently promoted the development of infographic illustrated Good Practice Manuals for the Utilization of Buildings of the UP. The Manuals were distributed in 20.000 printed copies in the 14 Faculties of the University of Porto. A Digital version was disclosure trough the University website as well as in each Faculty website.

These Manuals intend to raise the awareness and responsibility of users, as well as to enhance the potential of involving them in PPC, by providing guidelines on with (1) Precautions, (2) Recommendations and (3) Emergency occurrences (Figure 2) in a language that is accessible to all. Hence, users are a key actor in PPC, helping to avoid improper use, preventing risky situations, contributing to the recording of information and collaborating in daily maintenance activities (Ferreira 2014a).

Figure 2. Good Practice Manual for the Utilization of Buildings of the University of Porto (T. C. Ferreira, R.C. Rodrigues, P.F. Rocha; UP: R. J. G. Ramos; design: I. Vieira, S. Ruivo).

4 FACULTY OF ARCHITECTURE (FAUP)

4.1 *Context*

The complex of the Faculty of Architecture of the University of Porto (FAUP) was designed by Álvaro Siza, a former Professor and the winner of the Pritzker Prize in 1992. The buildings were constructed (1985-1993) in the Polo III of the University of Porto (Campo Alegre), bordered to the north by one of the main access roads to the city of Porto, to the east by the former XIXth century house of Quinta da

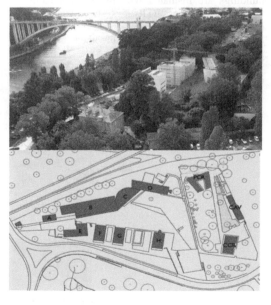

Figure. 3. Aerial view and Plan of FAUP. Main Buildings (A. Bar; B. Administration and Auditorium; C. Exhibitions; D. Library; E, F, G, H. Classrooms and offices); PCR – Carlos Ramos Pavilion – Classrooms; CAV – Old Stables (Polyvalent areas); CCR – Casa Cor de Rosa (Research Centre - CEAU).

Póvoa (Casa Cor de Rosa) and garden, which is located at the highest point of the terrain, and to the south by terraces over the River Douro.

The first works to be performed by Alvaro Siza, in 1985-1987, were the reconversion of the former House and Stables in services and classrooms and the construction of the Carlos Ramos Pavilion, an iconic piece of Portuguese Architecture with strong relation with the pre-existent garden in the upper platform of the site (Siza et al, 2003).

Afterwards (1986-1993), Siza completed the construction of the main buildings of the Faculty (A to H): a long building with central services protecting the courtyard from the highway and four towers with classrooms and offices facing the river. The buildings of FAUP are currently listed as buildings with a special interest in the Heritage Charter of the Porto Municipal Master Plan and are inventoried in the XXth Architecture Survey (Ordem dos Arquitectos, 2006).

4.2 *Management*

The FAUP buildings were initially designed to house 500 students, but are currently used by about 1100 students. The building has thus been forced to adapt to a more intensive use (24 hours per day and 7 days per week).

To better understand and diagnosis the current situation regarding PPC interviews were performed to managers, maintenance and cleaning staff and users (Ferreira 2017a).

The body responsible for the maintenance of FAUP buildings is the Executive Board, with one of its members being responsible for Premises and Equipment, a portfolio that includes such different aspects as cleaning, procurement of equipment and furniture, networks (water, electricity, computers, etc.), services and facilities, everyday maintenance and repair work, as well as occasional interventions requiring more profound conservation works.

According to those responsible, the main difficulties in managing and maintaining FAUP buildings are economic and administrative in nature, either because of the limited funds made available (which necessarily results in a shortage of human resources) or because of the administrative constraints that often affect the flexibility and speed of processes.

Nevertheless, these conditions have been improving gradually, in terms of both the internal organisation and the level of support provided by the Rectory of the University of Porto (UP), with more comprehensive interventions being undertaken since 2014.

The everyday users of FAUP spaces are employees, teachers, researchers, students and visitors. It is also important to emphasise the role of the security guard at FAUP: in addition to managing the keys to the rooms for teachers and students, they assist in the supervision of all spaces, recording any occurrences and damages that are then reported to the FAUP administrative services.

4.3 Maintenance

The day-to-day maintenance of FAUP buildings is carried out by three operational technicians, who are part of the Faculty staff and perform routine maintenance activities (gardening, cleaning of exterior spaces, inspection of roofs, checking and minor repairs in electricity, plumbing, networks, installations and furniture, among others), as well as giving support to the installation of exhibitions and other events held at FAUP's premises (Figure 4). In addition to performing everyday tasks, the operational technicians also execute other more profound tasks during school breaks, especially in July and August, such as the general checking of the electrical equipment, the painting of interior spaces, the repair of furniture, window frames and other important elements for the proper functioning of the premises at the beginning of the school year.

The cleaning of the interior spaces is outsourced to an external company according to specifications that establish the procedures and the products to be used. Currently, this cleaning work is performed by ten employees, who clean FAUP's interior spaces (from 6:30 to 8:30 am), including the sanitary facilities, classrooms, corridors, library, auditoria and administrative services. The windows are cleaned twice a week (amounting to 4 hours work per week). Every year, during the summer holiday period, the interior spaces (floors, walls, stairs) are cleaned more thoroughly, and the wooden flooring is varnished and sealed after having first been sanded with a machine. Twice a year, the outside windows to which access is difficult are cleaned by workmen using hoists and an elevated platform.

Although all the staff involved is extremely devoted to maintenance and cleaning tasks many of the actions are reactive, and the interviewed managers, users and staff admitted that having a systematic PPC tool would be of major utility and importance.

4.4 Constructive characterization

Regarding the construction techniques and materials, there are different solutions in the pre-existent and new buildings of the complex. Casa Cor de Rosa has load bearing walls in granite stone masonry coated with plaster, wooden frames in doors and windows, wooden structure in floors and roof, covered with Marseille ceramic tiles. The Old Stables have also masonry walls, wooden and concrete horizontal structures, covered with ceramic tiles.

At the Carlos Ramos Pavilion, there is an inverted flat roof, composed of a light concrete slab, an asphalt membrane for waterproofing purposes, thermal insulation in extruded polystyrene sheets, a geotextile blanket and a heavy protective layer of gravel. The walls are built with an experimental solution of brick wall and concrete pillars with external thermal insulation coating, while window frames are in steel with single glazing.

As far as the main buildings of FAUP are concerned, the constructive system consists of load-bearing walls and slabs in reinforced concrete. The roofs of the main buildings (A, B and C) and the towers (F to H) are covered with a zinc standing seam system, consisting of a light concrete slab, thermal insulation of black cork agglomerate and a zinc sheet cladding. The exterior walls of the main buildings are insulated with thermal insulation on the outside in an "ETICS" system, consisting of expanded polystyrene sheets glued to the reinforced concrete wall, previously smoothed and waterproofed (with "ceresite") (Figure 5).

The last layer consists of thin plaster based on the use of acrylic mortars, reinforced with a fibreglass

Figure 4. Regular inspection, maintenance and cleaning (Daniela Silva, the author, Clara Vale).

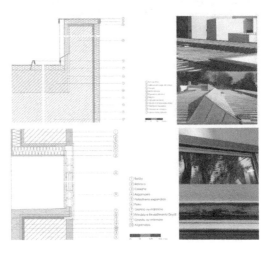

Figure 5. Details and photos of coverings and windows of the main buildings (Daniela Silva).

net, and applied in multiple layers. The finish used at the base of the buildings consists of exterior protection panels made of granite and limestone.

5 RECENT WORKS IN FAUP (2014-2018)

5.1 *Previous condition*

Broadly speaking, the main problems and alterations at FAUP stem from the natural wear and tear of a public building and the intensive use for around 25-30 years with no comprehensive conservation works.

Among the main causes of the degradation of the buildings are its exposure to atmospheric agents and the deterioration of materials and construction systems, which leads to infiltrations, particularly in the roofs and sills. Casa Cor de Rosa, the main buildings of FAUP (A to H) and the Carlos Ramos Pavilion had several anomalies in roofs, window frames, interior pavements, bathrooms, among others. The Old Stables also present several anomalies related to the lack of maintenance and will be the next building to be preserved, predictably in 2020-21 (Figure 5, 6).

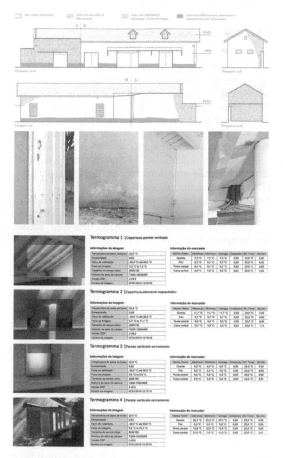

Figure 6. Decay Mapping and thermography monitoring of Old Stables (Eleonora Fantini, the author).

Generally speaking, the exterior walls of the buildings, before the conservation works, presented various occurrences such as stains, biological colonisation, cracks, blisters and deformations in the coatings of plasters and ETICS system. As far as the coverings are concerned, the cork agglomerate was rotting and the zinc flashing was becoming deformed, which has led to some infiltrations. Moreover, the main anomaly identified in the exterior window frames was the wear and tear and detachment of the paint, with some areas presenting signs of iron corrosion. Also, in the exterior pavements, some concrete slabs had cracked, and that the microcubes had become detached from the pavement due to the growth of the roots of the surrounding trees.

Regarding the problems related to the building's intensive use, it was also possible to identify some occurrences that are commonly found in a school of architecture where lessons are frequently of a practical nature, namely: wear and tear and the general deterioration of classroom furniture (benches and drawing tables, breakdown of equipment such as heaters and lamps, damage or malfunctions in the sanitary installations, dirt or writing on the walls and windows, accumulation of litter and rubbish in the classrooms and outdoor spaces.

5.2 *Intervention*

Since 2014, conservation works have been performed in the different buildings of the Faculty with the general coordination of Álvaro Siza. The first works occurred at Casa Cor de Rosa and included the conservation of the wooden structure, the insulation of the roof to provide comfort for offices in the attic, introduction of new waterproof membrane and tiles, renovation of bathrooms and kitchen, conservation of internal and external plasters and pavements, improvement of btechnical installations (heating, wireless, water and electricity) [1].

In 2016-2018, conservation works were performed in the external envelope of the main buildings of FAUP (A to H), being funded by the Rectory of the University of Porto [2].

The solution that has been adopted in the roofs is to apply water repellent paint to the light concrete slab, to replace the cork agglomerate with extruded polystyrene and to finish with the application of a studded rubber waterproof membrane, over which the zinc sheet is then laid. At the same time, the intervention is expected to include the replacement of the existing zinc flashing with a new one, with a slightly different profile, which prevents the water from seeping into the walls.

As far as the external façades are concerned, different incidences and types of anomalies were mapped allowing for the recovery of the façades, corresponding to the different levels of intervention: (1) areas without any anomalies, where the application of a new coat of paint is the only procedure that is required; (2) areas of superficial anomalies, such as stains resulting from biological colonisation – in this case it is recommended that a fungicide is applied, followed by a new coat of paint, after washing the surface with water applied at low pressure; (3) areas with detachments of the surface layer, requiring only the replacement of the final coating layer and waterproofing ("Visoplast"); (4) zones with deeper blisters, cracks and detachments which require the replacement of the various layers of thin plaster; in some cases, deep damages justify the replacement of the entire "ETICS" system (including the expanded polystyrene sheets) (Figure 7, 8).

Also, window frames were freshly painted both on the outside and the inside, including the replacement of the parts that are necessary to ensure the windows' proper functioning.

More recently (2018), works were carried out at Carlos Ramos Pavilion aiming at its conservation accordingly with the original design, such as the replacement of the roof covering following the original solution, the treatment and painting of external walls, the conservation of window steel frames with new glass with UV protection, replacement of the linoleum floor, bathroom renovation. Also, some updates and

Figure 9. Carlos Ramos Pavilion after the works (the author).

comfort improvements were performed in installations and devises, namely a new plasterboard wall attached to the pre-existent with new electric, communication and heating installations, new drainage system for bathrooms, new distribution of luminaires to adapt to a more flexible use of interior spaces [3]. (Figure 9).

5.3 Economic perspective

The record of the costs of the past works in the Faculty since it was built in 1985, together with data form recent interventions, allowed to develop a speculative chart comparing a curve of estimated costs of those works, with what it could have been a more virtuous curve with PPC (Figure 10). The latter is a schematic curve because there would also be spikes for cyclical maintenance over this period.

This chart intends to illustrate, in conceptual terms, the economic advantages of PPC strategies. Hence, besides the relevant cost reduction with cyclical maintenance, it is easier to get smaller funding for preventive routine works, than higher amounts for deep reactive intervention. Also, data from recent interventions (lifespan of materials and techniques, maintenance requirements, executive procedures,

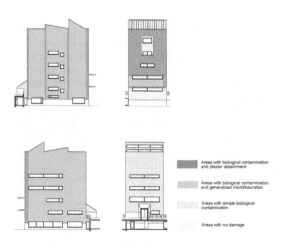

Figure 7, 8. Degrees of intervention in the façades of the main buildings (CEFA, FAUP). Photos of the main buildings during and after the works (Luís Ferreira Alves).

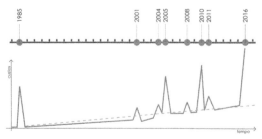

Figure 10. Speculative schematic chart comparing a curve of estimated costs of those works, with what it could have been a more virtuous curve with planned maintenance works in dashed line. (Daniela Silva).

costs) will allow to schedule and estimate more accurately the future costs on cyclical maintenance.

Nevertheless, it's important to remark that administrative and financial constraints (rules for public contracts, heavy bureaucracy, difficulty of anticipation of funding for prevention/maintenance, among others) make it difficult to proceed to the implementation of PPC in public buildings such as FAUP. Moreover, in Portugal the PPC legal framework and instruments are still poorly defined, and there are no funding or incentives for this kind of strategies (Ferreira 2014a).

6 CONTRIBUTIONS FOR PPC IN FAUP

The previous characterisation and diagnosis, as well as the data collected from recent works, allowed for the definition of some strategies and contributions for improving the maintenance and use on FAUP's buildings.

With this purpose, exploratory applied research was developed through the digital database computerized software and App MPLAN (Figure 11). This has permitted the compilation of different forms for each building: (I) Identification of Building and Users, (II) Characterization and Chronology, (III) Inspection and Diagnosis, (IV) Maintenance and (V) Utilization Manuals. These forms identify the users, techniques and materials, anomalies and, consequently, establish a plan PPC actions (inspection, cleaning, repair, etc.) scheduled according to their periodicity (daily, weekly, monthly, annual, multiannual). However, in addition to the preventive or routine maintenance actions, there are always unforeseen events that require actions of an exceptional or urgent nature, which must be undertaken in a timely manner and provided for by a specifically directed reserve fund.

Broadly speaking, current maintenance actions in FAUP may include, among others: cleaning and treatment of the gardens and outdoor spaces; inspection and cleaning of roofs and rainwater drainage systems; inspection and general adjustment of window frames; checking and adjustment of furniture; painting of interior walls; application of antifungal products, treatment and painting of exterior coatings; checking and inspection of the electricity system and other networks, infrastructures and equipment, according to the legislation and the specific procedures in each case, among others.

It should also be noted that any planning or intervention work to be carried out requires the previous authorization of the author (Alvaro Siza) and the Executive Board of FAUP.

On the other hand, since maintenance is only possible with the participation of all the building users, it is essential to promote strategies for their involvement and empowerment. In this context, teachers play a key role at FAUP in raising awareness of good practices in the use of the space, particularly at the beginning of each academic year.

Fostering the engagement of users, illustrated Manuals were developed, namely an infographic poster with pictures elucidating some recommendations for the use of spaces (Figure 12). The poster contains three levels of recommendations, from (1) Suggestions (organisation and cleaning of classrooms, furniture protection, recycling), (2) Alerts (stowage of scale models, electrical wiring and equipment care) and (3) Prohibitions/Obligations (rules for the use of the sanitary facilities, the placement of litter and rubbish in their appropriate bins, writing or sticking posters on the walls and windows) (Silva 2016; Ferreira 2017b).

At the same time, strategies can also be implemented that reward those classes that best exemplify good practices in the use and cleaning of spaces. This initiative can be an incentive to increase awareness of the good use and preservation of spaces, until the time comes when such care is no longer a rule and becomes an everyday and automatic habit for all users (Silva 2016).

Finally, among other suggestions, several interviewees expressed an interest in creating a participatory system for signalling occurrences in

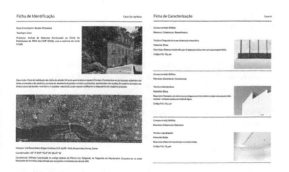

Figure 11. MPLAN Identification (Casa Cor de Rosa) and Characterization Forms (Main Building, Tower H) (the author).

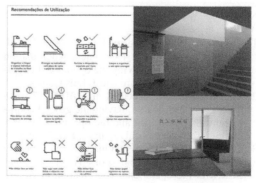

Figure 12. Poster with recommendations for users (Inês Vieira, Teresa Ferreira, Daniela Silva).

the building online (for example, through "sigarra", the Faculty's website with an institutional login), accessible to students, teachers, employees, operational technicians and cleaning staff. This could generate notifications addressed to the appropriate officials, technicians and members of the Faculty's Management Board responsible for the maintenance of the premises and its equipment.

7 CONCLUSIONS

This research demonstrates that FAUP buildings have been performing well, with good levels of satisfaction by users. Most of the damages are related to the natural ageing process and the intensive use of a school building, and were partially resolved in recent works in the buildings (2014-2018). However, currently, maintenance actions are still often performed in a reactive base (post-damage), leading to the dispersion of human and material resources, as well as to damages in buildings and infrastructures.

In this way, the implementation of PPC strategies (pre-damage) with support on a systematic methodology through MPLAN database computerized tool can be a good strategy for improving management and maintenance at FAUP. This approach will bring benefits such as cost-efficiency, updated management of information, decreasing of damage and risks, improving building's performance, as well as ensuring the preservation of the building's authenticity and integrity. Also, records from recent intervention provide important data for future PPC strategies. Nevertheless, it's important to remark that the administrative procedures and financial constraints in Portugal make it difficult to proceed to the implementation of PPC in public buildings such as FAUP.

Finally, this research intends to bridge the 'material component' with the 'intangible component', namely through the development of Users Manuals for wide diffusion, underlying the importance of involving and empowering all users in the PPC for built heritage.

NOTES

1. The works were performed in 2014/2015 with a cost of 148.316.09. The general coordination was of Álvaro Siza, having the operational coordination of Nuno Valentim and J. Luís Gomes.
2. The works were executed in 2016/2017 with a budget of 395,062.95, having the general coordination of Álvaro Siza and the operational coordination of Eliseu Gonçalves and J. Luís Gomes.
3. The works were performed in 2018 with a budget of 283.000. The general coordination of Álvaro Siza with the operational coordination of Álvaro Fonseca and J. Luís Gomes).

ACKNOWLEDGEMENTS

Assignment co-financed by the ERDF through the COMPETE 2020 - POCI and national funds by the FCT POCI-01-0145-FEDER-007744.

REFERENCES

Canziani, A. (ed.). 2009. *Conserving Architecture: Planned Conservation of XX Century.* Milan: Electa.

Della Torre, S. et al. 2003. *La conservazione programmata del patrimonio storico architettonico. Linee guida per il piano di manutenzione e il consuntivo scientifico.* Milano: Guerini.

Ferreira, T.C. 2014a. Towards maintenance: concepts and Portuguese experiences. In R. Amoêda, S. Lira & C. Pinheiro. *REHAB 2014*, Tomar, 19-21 March 2014. Barcelos: Green lines Instituto para o Desenvolvimento Sustentável.

Ferreira T.C., 2014b, Some considerations on the preservation of 20th-century architectural heritage. In Riso, Vicenzo (ed.). *Modern building reuse: documentation, maintenance, recovery and renewal.* Guimarães: EAUM.

Ferreira, T. C. 2017a. Faculdade de Arquitectura da Universidade do Porto. In T.C. Ferreira & P.F. Rocha (eds). *Saber Manter os Edifícios: pensar, desenhar, construir. Edifícios da Faculdade de Arquitectura e da Faculdade de Engenharia da Universidade do Porto.* Porto: CEAU-FAUP e CEES-FEUP/Afrontamento.

Ferreira, T. C. 2017b. Contributions for the implementation of preventive conservation and maintenance strategies in the Faculty of Architecture of the University of Porto. In R. Amoêda, S. Lira & C. Pinheiro (eds). *REHAB 2017*, Braga, 14-16 July 2017. Barcelos: Green lines Instituto para o Desenvolvimento Sustentável.

Ferreira, T.C. 2018a. Bridging planned conservation and community empowerment. *Journal of Cultural Heritage Management and Sustainable Development* vol. 8 (n°2): 179–193.

Ferreira, T.C. 2018b. Sustainable strategies in built heritage preservation: planned conservation and community participation. Experiences in Portugal. In K. Van Balen. & A. Vandersande (eds.). *Innovative Built Heritage Models, Reflections on Cultural Heritage Theories and Practices A series by the Raymond Lemaire International Centre for Conservation.* vol.3. Leuven: Taylor & Francis.

Ordem dos Arquitectos (ed.). 2006. *Inquérito à arquitectura do século XX em Portugal.* Lisboa: Ordem dos Arquitectos.

Ramos, 2017. A Vida dos edifícios: a manutenção do património edificado da Universidade dp Porto. In T. C. Ferreira & P.F. Rocha (eds) *Saber Manter os Edifícios: pensar, desenhar, construir. Edifícios da Faculdade de Arquitectura e da Faculdade de Engenharia da Universidade do Porto.* Porto: CEAU-FAUP e CEES-FEUP/Afrontamento.

Silva, D. 2016. *Perspectivas para a manutenção e utilização do Pavilhão Carlos Ramos na FAUP.* Master's Dissertation, Faculdade de Arquitectura da Universidade do Porto.

Siza, A. Et al. (Coord.). 2003. *Edifício da Faculdade de Arquitectura da Universidade do Porto. Percursos do Projecto.* Porto: Publicações Faup.

Siza, A. 2017. Interview by T. C. Ferreira. In Ferreira, T.C. & Rocha, P.F. (eds). 2017. *Saber Manter os Edifícios: pensar, desenhar, construir. Edifícios da Faculdade de Arquitectura e da Faculdade de Engenharia da Universidade do Porto*. Porto: CEAU-FAUP e CEES-FEUP /Afrontamento.

Van Balen, K. & Vandesande, A. (Eds.) 2013. *Reflections on Preventive Conservation, Maintenance and Monitoring of Monuments and Sites by the PRECOMOS UNESCO Chair*. Leuven: Acco.

Vandesande, A. et al, Preventive and planned conservation as a new management approach for built heritage: from a physical health check to empowering communities and activating (lost) traditions for local sustainable development. *Journal of Cultural Heritage Management and Sustainable Development* vol. 8 (n°2): 78–81.

Preventive Conservation - From Climate and Damage Monitoring to a Systemic and Integrated Approach – Vandesande, Verstrynge & Van Balen (eds)
© 2020 Taylor & Francis Group, London, ISBN 978-0-367-43548-6

Preventive monitoring and study of insect damage of carpenter bees to timber components of Chinese historic buildings

Y. Gao & Y. Chen
School of Architecture and Urban Planning, Shandong Jianzhu University, Jinan, China

D. Xu
Shandong Conservation Engineering Institute of Vernacular Heritage, Jinan, China

E. Li & J. Li
Xingtai Municipal Cultural Heritage Office, Hebei Province, China

Z. Ge & Y. Zhou
College of Information and Electrical Engineering, Shandong Jianzhu University, Jinan, China

ABSTRACT: The insect damage is one of the most harmful and widely spread damages to timber historic buildings. This paper illustrates a research program on the insect damage of carpenter bees led by a research team at the National Key Lab of Vernacular Heritage Conservation by Chinese State Administration of Cultural Heritage (SACH) since 2013. It adopted an approach combining the field monitoring and laboratory monitoring that conducted a full-year documentation and study on the different types of infestation on the wooden components. The research has collected a relatively large amount of basic data, which merit the study as a baseline for the preventive conservation of the insect damage of carpenter bees to wooden components of Chinese historic buildings.

1 INTRODUCTION

Since May 2013, a joint research team at the National Key Lab of Vernacular Heritage Conservation by Chinese State Administration of Cultural Heritage (SACH) has launched a research program on the insect damage of carpenter bees which adopted an approach combining the field monitoring and laboratory monitoring. It conducted a full-year documentation and case study on the different types of infestation on the wooden components, such as burrowing, nesting, spawning, and reproduction of different species in different seasons and environments. Up to the present, this research has collected a relatively large amount of basic data, which merits the study as a baseline for the preventive conservation of the insect damage of carpenter bees to wooden members of Chinese historic buildings.

2 DESCRIPTION OF THE RESEARCH

2.1 Background

The opportunity of this research originated from the repair of the front hall of Tianning Temple, one of the national listed historic buildings of Xingtai City, Hebei Province in 2014. Commonly known as Xida Temple, Tianning Temple, according to *Xingtai County Records*, was first built in the Tang Dynasty (618- 907A.D.) and was once a grand Buddhist temple. After the Chinese Communist Party came into power in 1949, the gate house and the main hall of the temple were demolished. At present, only the front hall remains (Figure 1). The dating analysis of the existing front hall completed by researchers of the Institute of Ancient Architecture of Hebei Province indicates it was built in the Ming Dynasty (1368-1644 A.D.). Given the small volume, the large size of the column was rare in the Ming Dynasty and the arches were complicated in types and divers in forms (Lin, 1999). In the process of repairing the front hall of Tianning Temple, it was found in the corner bucket arches the evident existence of the cavities (Figure 2), and multiple cavities in the corner arches continue to increase. After the conservation workers used small wooden sticks to block the cavities, large wasps that frequently appear around the building attracted the attention of the staff. It was therefore inferred that the phenomenon of insects in the front hall of Tianning Temple should be closely related to such bees.

It is well known that the soluble sugar, starch, cellulose, hemicellulose, protein, fat, minerals, etc. contained in the wood provide the nutrients needed for the growth of most wood pest larvae, and the wood is highly susceptible to insect pests (Yang, Chen & Liu 2009). This biological characteristic of wood makes

Figure 1. Front hall of Tianning Temple ©LIU Wanting.

Figure 2. Cavities in the corner bucket arch group of the front hall of Tianning Temple ©LIU Wanting.

Chinese traditional historical buildings with timber frame as the main load-bearing structure frequently infested by insect pests. The surface symptoms of insects' damage to wooden construction are in various forms, such as surface wormhole in a depth of less than one centimeter, worm path in the shape of flower twig under the bark caused by wood pests of the moth family, the pink wormhole caused by wood pests of lyctidae billberg and bostrichodea, the irregular tunnels and furrows caused by termites, and the large insect holes caused by beetle and bees (Chen, 2007). Among many wood pests, the carpenter bees like to dig holes in dry wood (Zhang, Wang & Li 2015). Once the holes come to scale, they will cause serious harmful consequences. The front hall of the Tianning Temple, which has stood for hundreds of years, is obviously the ideal place for the carpenter bees to survive.

After confirming that the cavities on the wooden components of the front hall of Tianning Temple are caused by the carpenter bees, the pest control of these bees is facing challenges. Most research on the prevention and control of timber pests in China focuses on the more endangered termites, and scant attention has directed to the hazards caused by carpenter bees and the relevant controls. The research results on this regard remain very limited. A small number of references mention the hazards and control of carpenter

bees, but with only a few words. In the article "A Study on the Prevention and Control of Ancient Buildings in Wooden Structures in Jiuhua Mountain", HE Xiangbin mentioned that cavities attributed to wood wasps caused the wood fibers to break, greatly weakened the strength of the wood, and their constant proliferation made the wooden components vulnerable and buildings collapse. Removal of flowering plants can be taken to sever the wood wasps' reproductive conditions (He, 2013). In the article "Prevention and Control of Insect Hazards of Carpenter Bees to Overhead Bare Lead-coated Communication Cable", XU Yinglin pointed out that anti-bee paint can be applied to the surface of the damaged components to drive away the carpenter bees (Xu, 1975). Despite the summative analysis and control methods proposed by these studies regarding the insect damage of carpenter bees, these studies are not systematic and lack of in-depth analysis of the specific pathogenic mechanism and symptoms of the insect damage of carpenter bees. Given the limitation of current research on the insect damage of carpenter bees and especially to the historic buildings, the joint research team decided to carry out the field monitoring and laboratory simulation environmental monitoring of the insect damage of carpenter bees to wooden components of historic buildings.

2.2 Field monitoring

Monitoring work is an important part of the preventive conservation of historic buildings. It is a series of processes for collecting observations considering the purpose for special control and management (Wijesuriya, Wright, & Ross 2004), and a process of conducting systematic and continuous collection and analysis of each going-on task. (Abbot and Guijt, 1997). The monitoring work over carpenter bees is a rather long process and can be divided into two phases: the initial phase of field monitoring and the later phase of laboratory monitoring. In the field monitoring phase, after careful observation and data comparison, the subspecies of the carpenter bees distributed in Hebei Province were mainly categorized into the yellow-breasted bees and the carpenter bees (Figure 3), and the latter was discovered this time as one of the subjects under the observation of its serious damage. The carpenter bee is an insect of xylocopidae in medium and large-size. Its body is thick and strong with the male body of 24-25 mm. long and the female body of 25-26 mm. long. The female bee has a short needle at the tail. Living alone, these bees like to build nests by digging channels and holes on the dry wood and are very harmful to wood, bridges and buildings (Wu, 1982).

The initial field monitoring work carried out from 2014 to 2015 focused on the phenomenon of insect pests of carpenter bees to the front hall of Tianning Temple. Soon after, the repairing materials preparation workshop behind the front hall of Tianning Temple was also found the traces of damages left by carpenter bees on timbers, including timbers that

Figure 3. Specimen of male and female carpenter bees ©LIU Wanting.

have been processed into sheets (Figure 4), bucket arch components, and rough logs. The joint research team then expanded the scope of the monitoring to all buildings in Tianning Temple and accessed to a lot of valuable first-hand field research data. Combining the data collected in the field monitoring with the literature research, we found that:

1) The damage behavior of carpenter bees on timber components of ancient buildings is different from that of common Lyctus spp. The carpenter bees mainly leave cavities by gnawing at timber components, which serve as the nests for their larvae to grow. But the larvae do not feed on the timber. 2) The destructive behavior of the wood wasps is found more serious on timber components made of elm and pine. This is related to the timber properties of these trees. In the north China, elm and pine trees are not resistant to cavities (Zhou & Chen, 1989). 3) Timber components made of wood with low hardness and high sugar content are more susceptible to such insect pests.

2.3 Laboratory monitoring

After a period of field monitoring, the research team realized that it was necessary to conduct special laboratory simulation environmental monitoring on carpenter bees to improve the efficiency of

Figure 4. Section cutting plane of insect pests of carpenter bees in material preparation workshop ©LIU Wanting.

monitoring and obtain more intuitive and controllable results. Laboratory monitoring conducted between 2016 and 2018 is divided into two phases: the first phase is indoor incubator monitoring and the second phase is outdoor laboratory monitoring.

In the first phase, the researchers trapped several carpenter bees in a glass incubator on September 4 of 2016 for observation, collecting luffa flowers, and setting up drinking water to provide the survival conditions for these bees. However, the experiment failed. In many attempts, the bees died in a short period of time, and the survival duration was too short to meet the observation needs. At the same time, the researchers set up another experimental device for indoor monitoring: the carpentry workshop made of wood frames covered with gauze, within which multiple logs of elms and pines as renovation materials were placed there. During the indoor laboratory monitoring process at this phase, although it was not fully successful, it helped to obtain a lot of important data information. The observation team found that:

1) The carpenter bee has strict requirements on the living environment, and it is difficult for it to survive for a long time if only the living conditions for food is satisfied. 2) When a carpenter bee is threatened by the outside world, it will eject liquid from the mouthparts to intimidate the enemy, and some will stick out poisonous needles to attack the enemy. 3) In the indoor carpentry workshop monitoring device, the observation team found eleven timber pests (including the carpenter bee) and some of them were suspected to be braconid.

The monitoring of indoor laboratory failed to obtain effective observation results for the destructive behavior of carpenter bees, largely because the living environment could not meet the needs of carpenter bees. To further understand the destructive behavior of the carpenter bees and their larvae inside the timber and assess its damage, the research team decided to start the second phase of outdoor laboratory monitoring: simulating the outdoor living environment of carpenter bees by building a glass house in wooden frame covered by metal gauze to ensure air circulation. Meanwhile, luffas were planted within the glass house to provide flowers as a source of food for the carpenter bees. In the selection of wood, elm and pine wood, which were most vulnerable to insect pests among the building repairing preparation materials, were selected as the experimental materials. Many carpenter bees were trapped and put into the glass house. After more than one year of monitoring and observation, it was found that the carpenter bees lived well. The first batch of timber was found with many symptoms of insect pests caused by carpenter bees. In addition, the unfinished cavities of carpenter bees were also found on the wood frames. The monitoring at this stage has led to a few important conclusions:

1) The destructive behavior of the carpenter bees is mainly manifested in spawning in holes on the

surface of the timber component. The hole is about 7-8 mm in diameter and in relatively regular shape. The diameter of the inner tunnel is slightly larger than that on the outer surface, and the cross section of the tunnel takes the shape of an irregular curved line (Figure 5). A carpenter bee lays several eggs in the tunnel, and each larva is covered with leaves and with beeswax. The opening is blocked with beeswax or a mixture of clay beeswax.

2) In Xingtai area of Hebei Province, the carpenter bee begins to show activity marks when the temperature is above 15 degrees Celsius. In addition, the monitoring team unexpectedly found that the carpenter bees have winter dormancy behavior.

3) The carpenter bee also chooses the existing holes in the timber components for breeding.

4) The carpenter bee has certain requirements for the depth of the cavities for breeding. When the size of the hole does not meet the depth of its cavities, the carpenter bee will give up the created hole.

5) The monitoring team confirmed the severity of the damage to the timber parts caused by the carpenter bees, and some of the experimental materials showed a wormhole tunnel running through the radial section (Figure 5). According to the habit of the carpenter bee, if the nest cannot be extended to meet the need for the depth of the cavity, the carpenter bee usually redirects the hole and re-gnaws the hole. If such insect pest continues, it will seriously affect the mechanical properties of the timber component.

The successful monitoring of the first batch of timber enables the researchers to gain a certain understanding of the formation of the wormhole of carpenter bee and the living environment of the larvae. At the same time, the use of existing cavities for spawning by carpenter bees also triggered researchers' interest in exploring the survival conditions of larvae. The research team then set up a second set of experimental devices, namely, manually drilling the hole simulating

Figure 5. Radial section cutting plane of cavities created by carpenter bees ©LIU Wanting.

Figure 6. Outdoor experimental device simulating the cavities ©LIU Wanting.

the common cavities of carpenter bee on the pine board and setting the duct at the back and applying a black cotton cover (Figure 6). However, in the long-term monitoring, no carpenter bee chose to spawn in the experimental device. Despite the failure of the experiment, the research team concluded that the carpenter bee made its own choices regarding the existing holes on the wood and the environment inside the hole is one of the important factors for consideration in its breeding activities.

3 CONCLUSIONS

Based on the discussions in the previous sections, the main conclusions can be drawn as following:

This study, combined with field monitoring and laboratory monitoring, basically reveals the specific pathogenic mechanism and symptoms of insect pests caused by carpenter bees to timber components of historical buildings, and confirms the specific effects on the mechanical properties of timber components that carpenter bees caused by breeding in the tunnels they gnawed. This study fills the gap in China's domestic research on the insect pests of carpenter bees and lays the foundation for the preventive protection of insect pests to timber components of China's historical buildings caused by carpenter bees.

The field monitoring identifies the breadth and severity of the damage of carpenter bees. In the laboratory simulation environment monitoring, it was confirmed that the survival condition of the carpenter bees in Xingtai area of Hebei Province and the active temperature was above 15 degrees Celsius, Furthermore, the carpenter bees were unexpectedly found to have the dormant period, which supplemented the important baseline research data on the China's insect pests of carpenter bees.

It was found that after screening the existing holes and their internal environment, carpenter bees may choose to use them for a nest for raising larvae. This finding also supplements the basic information on the breeding behavior of the carpenter bees.

The next experiment that the joint research team is going to conduct will focus on: 1) Studying which killing method is most effective for controlling the insect pests of carpenter bees while keeping the effect on the structural performance of the timber components small. 2) Researching what kind of preventive protection measures can be taken to effectively deal with the insect pests of carpenter bees.

A more intensive interdisciplinary approach is a must for the further progress of this research program in the near future.

ACKNOWLEDGEMENTS

We are indebted to the staff of Xingtai Municipal Cultural Heritage Office for taking assistance of field monitoring for this project. And the special thanks also go to our research partners at China Forestry Science Academe in Beijing for their help and support.

REFERENCES

Abbot, J., & Guijt, I. 1997. *Changing Views on Change: a working paper on participatory monitoring of the environment*. London: International Institute for Environment and Development (IIED).

CHEN Yunshi. 2007. *Wood Structure of Historic Buildings and Preservation of Timber Cultural Heritage*. China Architecture and Building Press.

HE Xiangbin. 2013. A Study on the Prevention and Control of Ancient Buildings in Wooden Structures in Jiuhua Mountain. *China Science Research on Cultural Heritage*, Issue 04: 60–63.

LIN Xiuzhen. 1999. A Study on the Founding Year of the Front Hall of Tianning Temple in Xingtai City. *Annals of Cultural Heritage*, Issue 03: 42, 44–46.

Wijesuriya, G., Wright, E., & Ross, P. 2004. Cultural context, monitoring and management effectiveness (Role of monitoring and its application at national level) in Stovel, H. ed. *Monitoring World Heritage*. Paris: UNESCO World Heritage Centre and ICCROM, 70–75.

WU Yanru. 1982. Research on the species of Chinese carpenter bees and descriptions of new species (Hymenoptera, Apoidea). *Zoological Research*, Issue 02: 193–200.

XU Yinglin. 1975. Prevention and control of insect hazards of carpenter bees to overhead bare lead-coated communication cable. *Material Preservation*, Issue 06: 11–13.

YANG Jinwei, CHEN Xiaofeng & LIU Jiuli. 2009. A Preliminary Study on the Prevention and Control of Wooden Structure Pests and Damages in Ancient Buildings. *Silk Road*, Issue 18: 81–86.

ZHANG Zubin, WANG Yuxia & LI Gongchun. 2015. Types, distribution and damage symptoms of main pests in buildings in China. in Proceedings of the 2015 Annual Conference of the Insects Society of the Three Provinces of Central China (Hubei, Hunan, Henan), 225–231.

ZHOU Ming & CHEN Yunshi. 1989. *Erosion Resistance and Pest Control for Houses of Wood and Bamboo Structure and Furniture*. China Forestry Publishing House.

Preventive Conservation - From Climate and Damage Monitoring to a Systemic and Integrated Approach – Vandesande, Verstrynge & Van Balen (eds)
© *2020 Taylor & Francis Group, London, ISBN 978-0-367-43548-6*

Condition assessment and monitoring in Milan Cathedral: Putting risk assessment into practice

L. Cantini
Department ABC, Politecnico di Milano, Milan, Italy

F. Canali
Veneranda Fabbrica del Duomo, Milan Italy

A. Konsta & S. Della Torre
Department ABC, Politecnico di Milano, Milan, Italy

ABSTRACT: The cathedral of Milan required continuous care during the long realization works, since 1386, its foundation, to 1964, officially indicated as the end of the construction. The Veneranda Fabbrica del Duomo, the body managing the construction and the conservation of the Cathedral in the course of time, developed a series of technical competencies and procedures for this purpose. These good practices are also supported by the Politecnico di Milano. Among the various activities carried out by Politecnico di Milano for sustaining the Veneranda Fabbrica in its mission, the risk analysis related to the rich decorative sculptures became a new point demanding the setting of a proper evaluation methodology. By analyzing different approaches among risk and condition assessment, the authors present a study of a new procedure for detecting risk conditions through a set of indicators. The aim of this procedure is to define an index connected to the frequency of the inspection activities necessary for guaranteeing safety aspects concerning the building and the people attending its spaces.

1 INTRODUCTION

The renewal of the collaboration between the Veneranda Fabbrica del Duomo and Politecnico di Milano, started in 2015, confirms the joint effort to rationalize and integrate the conservation practices of the Milanese Cathedral. More specifically, the aim is to limit the isolated and episodic actions and pursue the search for an efficient strategy and organization of the whole activities in a continuous and long-term process.

Given the complexity of the construction – owing to its dimensions, its structural and material characteristics, the continuous interventions in the course of time, as well as the present environmental threats – a multi-disciplinary approach is required, according to the methodological strategy described in Van Balen & Verstrynge (2016). In this perspective, the preparation of the investigation program is carried out by a team of different specializations with the purpose to improve our level of knowledge of the cathedral structure. In fact, the ongoing monitoring and laboratory testing activities contribute to a better understanding of the deterioration mechanisms and decay evolution, the detection of the principal environmental factors and the evaluation of their effects on the state of conservation of the building. The cathedral is subjected to vertical tilting of masonry walls and spires, high concentration of stress in

some tie rods located at the intrados of the arches, problems connected to the durability of the treatments applied during the last century and major water infiltrations, which spoilt the masonry structures from inside. The paper will offer a short description of those topics, appointed by the Veneranda Fabbrica and investigated by the support of various research teams form Politecnico di Milano.

After a tragic event concerning safety conditions into ancient churches, also the risk assessment connected to the decorated surfaces of the monument became another point of interest for the research program. The episode occurred in Florence, during October 2017, when a piece of a corbel, in Santa Croce Basilica, broke down falling on a Spanish tourist from a high of about 35m, causing his immediate death. The Italian association of the ancient cathedrals (named AFI), grouping several boards of experts managing different aspects for conserving these buildings, worked on the setting of a new procedure for monitoring the decay of the surfaces of those monuments in order to calibrate a methodology for preventing risks and the recurrence of such episodes.

This work presents the adopted methodology for risk assessment set by AFI, with some observations matured by applying the analytical tool to some scenarios concerning the Milan Cathedral, considering also other experiences addressed to the definition of a reliable procedure for putting risk assessment into practice.

2 THE DUOMO OF MILAN AND THE ON-GOING RESEARCH ACTIVITIES

The realization of the Milan Cathedral knew several phases, characterized by periods of fairly constant construction activities, and times with a slowing down of the works due to specific historical events (like wars) or particular technological difficulties. As a result, the need for a constant care of the built parts of the cathedral has driven to a continuous training in facing problems connected to materials durability and structural failure. The Candoglia marble stones, with their characteristic light white and clear pink veins, compose the external coating of the lower main masonry walls and the very structure of the whole 135 spires. This material was also used for the sculptures and the decorations of the building, like reliefs and window frames. The architectural features of the cathedral, realized in Candoglia marble since the 14th century, exposed to the interaction with the environment, showed some limits in term of durability and the substitution of decayed stones became one of the common practice adopted by the workers of the Veneranda Fabbrica. Small portions of Candoglia marble, when presenting a bad state of conservation, were replaced by new Candoglia stones, suitably shaped by the stonecutters. This practice became a common measure and is used also nowadays.

When the realization of the Cathedral was concluded, officially by the inauguration of the new doors of the main façade in 1964, the above mentioned practice was maintained for the most damaged parts of the complex, but new conservation methodologies were also introduced for protecting the materials by using, for example, new chemical treatments. In this period, the architect of the Veneranda Fabbrica, engineer C. Ferrari da Passano, focused on several aspects requiring strategic interventions: from structural problems connected to the large crack pattern characterizing the pillars of the dome, to the calibration of the restoration phases of the artistic elements made of Candoglia marble. Since the second half of the 20th century, new monitoring campaigns carried out by academic institutions, like Politecnico di Milano, were introduced on different elements composing the complex. The long relationship between the Veneranda Fabbrica and Politecnico di Milano was recently renewed by an agreement. The scientific support provided by the various academic teams is described below.

The measurements carried out on specific structures (like pillars and spires), starting during the second half of the 20th century, by geomatics techniques and never interrupted, are now supported by modern sensors and a deeper knowledge of their geometry, acquired by applying new advanced survey techniques (Achille et al. 2018). New monitoring systems for static, dynamic and thermal analysis were recently applied to different structures of the complex for improving the analysis on its dynamic response. The historical data series recorded along the time allowed the continuous assessment of the structural problems of the building and the risk interpretation.

The structural health monitoring comprises the measurement of different types of physical quantities, both static and dynamic. The on-going work started with a complete check of the tensile condition into the tie-rods at the intrados of the arches in the naves and continued to some selected structures (Gentile & Canali 2018).

Moreover, the acoustic pressure and the acceleration on the windows glasses, caused by the public concerts in Piazza Duomo square, were measured through accelerometers installed on the glasses. Further experimentation in the laboratory, seeing that international standards of exposure limits do not exist, provided a first indication of a safety coefficient (Canali et al. 2018).

Regarding the degradation phenomena on the marble surfaces, an extensive diagnostic and monitoring activity revealed the worsening effects produced by acid rains and other climate and factors on the Candoglia marble (Figure 1). In detail, colorimetric measurements and on-site digital microscopy were performed in order to assess the evolution of the main degradation mechanisms, triggered by urban pollution, direct rainfall, and seasonal variations in temperature (Fermo et al. 2018).

Among the various aspects, the risk induced by the rich artistic repertory of the cathedral has now become another critical issue to be investigated. The local state of stress can produce severe damage on the marble decorations, causing the cracking of the Candoglia stones with further deformations and detachments followed by stones fragmentation (Figure 2). The dangerous consequences of eventual falling of materials, in the case of the Milan Cathedral, is limited by the frequent inspections carried out by the

Figure 1. Exposed Candoglia marble elements under worsening process.

Figure 2. Detail of a cracked stone over a stone arch.

workers of the Veneranda Fabbrica. Periodical visual inspections are a consolidated part of the so-called good practices set by the Veneranda Fabbrica and still proposed as the foundation of the management process used for detecting problems and setting design interventions. Recently, on behalf of the Italian association of the Fabbricerie (AFI), the Veneranda Fabbrica technical staff has worked hardly on the idea of an in depth evaluation of the risk connected to the rich decorations characterizing the Italian Cathedral.

3 HISTORICAL BUILDINGS AND RISK ANALYSIS: TOWARDS A MORE SYSTEMATIC APPROACH

The definition of risk, that is the degree to which loss is likely to occur (Stovel 1998), as a function of two principal components "hazard" (probability of an unwanted event) and "vulnerability" (magnitude of consequences amplified because of the recognised values) is widely acceptable at an international level in the field of built heritage conservation.

According to the scope of the analysis – depending on the number (one risk, combination of risks, all possible risks), and the kind of risks examined, the scale dimension of the analysis, the time considered, as well as the extent of integration of physical, cultural, social and economic aspects – the assessment methods vary. Moreover, the different methods determine also the type of activities to undertake for the collection of information.

By trying to make some general considerations, with regard to the scope of the analysis and the relative assessment methods, we can notice the prevalence of a fragmentary vision of the problems and their solution. As a matter of fact, there have been elaborated many methods for the analysis of the different risks to which built heritage is exposed. From air pollution and acid rain to natural hazards, and from inappropriate intervention and use to mass tourism – the list is very long (Camuffo 1997, ICOMOS 2000) – we can find a relevant number of studies and practice involving respectively numerous specialists and

skills from a variety of fields. Obviously, this kind of "separation" arises from the necessity to deepen the knowledge of the different problems and their solution and it cannot be undoubtedly considered as a stumbling block. The problem emerges from the difficulty to coordinate interdisciplinarity, to understand multifaceted problems, and to put together all the findings.

In this direction, important results have been reached on the large-scale analyses, for risks that concern extensive areas. The example of risk maps (related principally to natural hazards and climate factors both aggravated from the climate change) is representative of this category. The analyses carried out on this scale are indispensable firstly, for the awareness of the problems and secondly, for the definition of policies and strategies able to assist the implementation of preventive measures at different governance levels.

On the other hand, a large-scale analysis cannot permit, mainly for the unavailability of time and economic resources, a detailed assessment of the building vulnerability and consequently of its loss, in order to define priorities and address the necessary measures at a building level. To meet this objective, it is mandatory an appropriate analysis of risks on the building scale. But when this analysis is carried out, what kind of methods are employed and in which way the decisions are taken? The questions are simple, yet the answers are complicated. When conservation practice is coherent with its theory and its basic principles – the definition of conservation as an "action taken to prevent decay and manage change dynamically" (Feilden 1982) is commonly shared in the occidental culture – the reasoning and the analysis of the possible causes of deterioration and their effects become indispensable activities.

With respect to the first question, two categories could be distinguished. Within the frame of an overall management/conservation process in the long run, or else a conservation plan, risk assessment contributes to an effective preventive system where the decisions and all the activities are consistent, coordinated and planned (Della Torre 2018). On the other hand, when decisions concern separated activities and it is necessary a heritage-impact assessment, as a reaction to a specific proposal to do something, understanding and assessing the risks is essential in order to avoid or mitigate the adverse impacts on the significance of the heritage (Clark 2014). Thus, both approaches, proactive and reactive conservation, as described by Clark (2014), demand an appropriate analysis by taking into consideration all the factors influencing the complex relationship of the built heritage with its environment.

As for the methods used, we can notice that during the last twenty years, efforts have focused on the development of more coherent and clear procedures for risk assessment, recognising on the one hand, the positive results achieved from a risk management approach, consolidated as a practice, to museums preventive conservation (see Michalski 1990, Waller 1995, Fry at al. 2007, Forleo 2017) and on the other hand, the need for effective tools in order to normalize best practices. Representative cases are the Guidelines for Conservation Plan approved by the Lombardy Region (Regione Lombardia 2003), the European Cultural Heritage Identity Card EU-CHIC project that has developed criteria and indicators for risk assessment (eu-chic.eu. 2009), the development of advanced diagnostics protocols (Kioussi et al. 2013), as well as the application of Failure Mode and Effects Analysis by Monumentenwacht Noord-Brabant (Naldini et al. 2018).

A further contribution in this direction is the development of two European standards: the EN 16096:2012, that concerns the condition survey of immovable heritage and the EN 16853:2017, that is related to the entire conservation process.

The first one provides guidelines for recording and assessing the condition of built heritage by visual observations, together with simple measurements. The scope of the standard is to implement preventive conservation; or rather, as stated in the document, it acts as the basis for recommending preventive conservation, maintenance and immediate repairs and for defining the need for further diagnostics of damages.

The first draft of the standard in 2010 was rather confused. Despite the consideration of relevant aspects related to the evaluation of risks during the phase of assessment, the identification of the related measures was not based on the previous analysis or rather it was not based in an explicit way. Indeed, the activities of condition and risk assessment were not separated (in the document the words 'risk assessment' were not even mentioned) and the classification of measures was linked directly to the condition classes (Della Torre 2010).

The final version of the standard, approved in 2012, has resolved the previous ambiguities firstly by pointing out the difference between condition and risk assessment, and secondly by setting up a classification of urgency categories connected to the risk analysis. In detail, risk analysis should take into account the following elements as described in the standard:

1. probable causes and triggers of the recorded condition,
2. external actions affecting the components and components assessed as probable causes of damage,
3. expected variations in external actions,
4. probable consequences of the recorded condition,
5. probability that, or the speed at which, the consequence and further deterioration will occur,
6. need for additional investigations,
7. probability that further investigation will reveal hidden damage and the consequence of this damage if found,
8. probable effect on and for historic significance,

9. relationships between the components and other elements,
10. other external and environmental factors which may significantly affect conditions and their probability (flood, fire, seismic activity, landslide etc.),
11. urgency of measures.

Lastly, the overall recommendation grading is based both on condition and risk evaluation.

Moreover, with the EN 16853:2017 the risk evaluation becomes an important component of the conservation process, necessary for the identification, evaluation and selection of conservation options. In particular, risk assessment is regarded on the one hand, as an integral part of identification/ investigation/diagnosis phase, intended as prognosis. On the other hand, it gives rise to a series of risk-related issues concerning the project evaluation.

An important issue that emerges from these documents is the necessity of a gradual process in which the knowledge acquired at each step provides the basis for further action. If during an initial assessment, based on direct observation, simple measurements and analysis of the available information on the building, the problems are easily understood – damages, causes and potential risks can be detected – decisions about the need of measures can be taken. If the obtained information is not adequate, a detailed assessment based on further investigations is recommended in order to carry out an appropriate diagnosis/prognosis and define the necessary interventions.

This kind of gradual knowledge process is also a common approach to the documents concerning the structural assessment of the historic buildings (e.g. ICOMOS – ISCARSAH 2003, ISO 13822:2010, CIB Commission W023 2010). What is more, as pointed up by the three documents mentioned above, these two levels of assessment, preliminary and detailed, determine also two different ways of decision-making process. The former, based mainly on visual observations and historical analysis, is a qualitative approach, or in philosophical terms an inductive procedure. In contrast, the latter includes laboratory and in-situ tests, field measurements and monitoring, and therefore is mostly characterized by a quantitative approach, or else a deductive procedure.

The uncertainties inherent in historic buildings, due to their complexity (spatial and material variability, transformations, uncertainties regarding material characteristics, unknown influence of previous phenomena, etc.), call for a combined analysis of the information gained from each of them, in order to enhance the reliability of both procedures (ICOMOS – ISCARSAH 2003).

4 THE PARAMETRIC TOOL FOR RISK ASSESSMENT PROPOSED BY AFI

4.1 *Inspections and good practices*

Through their seasonal regular inspections, the Veneranda Fabbrica set a long-lasting reliable procedure for enhancing the safety conditions of the cathedral. The characteristics of the cladding/building material, the Candoglia marble, showed common decay processes like scaling, pulverization and formation of black crusts or organic deposits. At least from the XIX century, Veneranda Fabbrica chief Engineer are used to describe the interventions set for the damages involving the Candoglia stone elements by a detail colored legend applied on a 1:10 scale drawing of the area under control; and the Veneranda Fabbrica has then developed a procedure aimed at identifying the elements requiring a complete substitution and the ones to be maintained by sealing and repointing or to be partially reconstructed by tessellation. This approach is used also for the wide decorations realized in Candoglia stones characterizing the building facades.

The Veneranda Fabbrica is trying to set an even more analytic procedure for evaluating, during periodical inspections, the risks connected to the stone decorations, dialoguing with other institutions managing important historical cathedrals. This cooperation gives continuity to a long-lasting tradition of these on-going building yards (Opera della Primaziale pisana 2012).

The idea is to propose a qualitative and semi quantitative analysis of the stone surfaces organized in different steps. It is an attempt for classifying the risk. By considering the visible conditions of the element, the estimated durability of the material and the regularity of the inspections, a computation of these parameters could drive to a safety index. To this aim, a series of simulations and sensitivity analyses were carried out for evaluating the reliability of this method. Such an index could undoubtedly support decision making in order to prioritize the zones, where a conservation yard has to be scheduled, according to current procedures. However, in order to control the risk of local accidental failures of hanging decorations, this procedure is highly depending on the capability of early detection, the durability estimations, and an organization oriented to small local interventions, more than to works extended on large zones.

A preventive approach requires enhanced knowledge on decay mechanisms and their causes: therefore, the compatibility among the different inbuilt materials (marble, mortars, iron devices, resins, etc.) is under investigation. For instance, the polymeric materials introduced during the restoration interventions in the 1970s gave rise to various damage types related to compatibility and durability aspects; and, as well, to effective ignorance of how a new component can interact with the old monument. Protective chemical mixtures were used both in surface parts (mortar joints, marble consolidation, protection and painting) and within the structure (passivating coating of metal bars). While the negative effects on the external facades have been already investigated, a continuous monitoring of the structure is carried out in order to comprehend in depth the relationship between the long-term alteration of the resins and the environmental conditions (Della Torre & Cantini 2018). As for the effects induced by thermal solicitations, on the exterior surfaces and in the interiors, the thermal and hygrometric conditions are now regularly monitored. The indoor microclimate represents another source of risk, potentially producing a negative effect on the diversity of materials conserved in the interiors. The microclimatic analysis carried out in the naves of the cathedral has the aim to detect trends and peaks in the internal temperature and relative humidity distribution (Aste et al. 2019).

Sharing this knowledge, it would be possible to detect the first signs of decay processes, and to distinguish the risks, which require an immediate intervention, and the ones to be monitored and treated in the frame of the planned maintenance operations.

4.2 *The safety index evaluation procedure set by the Italian Association of the Fabbricerie (AFI)*

AFI association is experimenting an analytical procedure aiming the identification of a safety index that should represent an alarm for the risk conditions observed on specific architectural elements. The logic of the evaluation procedure is calibrated on the common organization of the inspections, based on the subdivision of the building elements into technological classes. For each class and sub-element, a description form is arranged for recording specific information.

The evaluation tool, an electronic sheet, includes 5 degrees of judgments (from excellent to very bad) to be indicated for the following base parameters:

1. general state of conservation of the element,
2. presence of new cracks,
3. already existing cracks (previously detected),
4. falling effects for the considered element,
5. characteristics (thin, large, passing through, etc.) of the new cracks appeared after the last inspection.

These main voices are connected to other parameters. The frequency of the inspections, expressed in years by introducing an inspection periodicity parameter (Ta), plays a crucial role in the evaluation, together with the maintenance activities, the reliable duration of the interventions and the respect of the planned conservation actions. Moreover, the level of knowledge determined using investigation techniques is an aspect included in the analysis. This knowledge can be defined as exhaustive, good and minimum.

The above-mentioned parameters are used for calculating 2 factors: the inspection period factor and the maintenance period factor. By specific formulas, 4 main results are determined:

1. The operating state of the element;
2. The alert level;
3. The new inspection periods depending from the alert level;
4. The date of the next inspection.

The aim of the analysis is to draw out a prediction of the worsening conditions identified during the periodical inspections in order to set two main strategies: increasing the control activities, by shorter intervals between the periodical inspections, and setting specific interventions on the damaged elements, anticipating the planned conservation activities, like repairing, sealing or strengthening interventions.

4.2.1 *The qualitative parameters of the procedure*

A simulation carried out on a decorated window of the cathedral allowed observing the results obtained by the first part of the protocol set by the analytical tool. Table 1 shows the correspondences between the 5 base parameters and the values associated to each judgment. These parameters are put in relationship with the periodicity used for the inspections (Ta), expressed in years. Considering the case of the Milan Cathedral, 6 months are considered the common interval between the inspections.

The first 5 parameters are balanced on two main conditions: the appearance of new cracks, not observed in previous inspections, and the eventual increasing of the evidence of the already documented cracks. The evidence of the problem and the further risk evaluation is deeply connected to the detection of the crack pattern and its evolution in time. According to the intentions of AFI, the tool is based on the recording activity carried out by a trained staff, coordinated by experts. Cracks represents a clear sign of risk that should be recognized by different subjects involved in the periodical inspections.

The parameter "general state of conservation" should provide more difficulties in the objective definition of the problems afflicting the element under control. It refers to a broad vision of the element and gives the possibility to collect information regarding alterations or decays on the materials. Even a trained staff could provide different interpretations examining the same element. In order to limit problems connected to subjective judgements, the analytical tool was clearly addressed to the identification of a peculiar decay category: the cracks. Moreover, when new cracks are observed, the inspector has to define the dangerous level, choosing among 5 alternatives, from negligible to very bad.

4.2.2 *The role of the continuous care approach*

After defining the first set of parameters described above, the procedure takes into account important aspects connected to the frequency used for controlling and carrying out a constant maintenance of the considered element. This section of the electronic sheet is composed by 5 parameters collected in Table 2.

The parameters considered in the second part of the analytical tool are time oriented and have the aim to emphasize the role of periodical maintenance actions as key measures for the preservation of the various building elements. The tool will penalize the lack of maintenance and the lack of respect of the scheduled interventions. This logic is based on the first parameter indicated in Table 2: the nominal life of the maintenance intervention. This parameter represents a key-point and its definition is related to the effectiveness of the interventions carried out on specific elements with peculiar problems. Actually, AFI proposed a conventional nominal life of 40 years on the base of empirical experience matured along the important inspections and interventions tradition constituting a fundamental knowledge heritage for the setting of the proposed procedure.

Table 1. Base parameters and corresponding definitions with the associated values for the analytical tool proposed by AFI.

Parameters	Definitions values				
	1	0.8	0.6	0.4	0.2
General State of conservation (a)	excellent	good	moderate	poor	very bad
Presence of new cracks (b)	NO	YES	-	-	-
Previously documented cracks (c)	absent	light	moderate	bad	very bad
Falling effects (d)	negligible	law	moderate	bad	very bad
New cracks damage level (L_d)	negligible	law	moderate	bad	very bad

Table 2. Parameters collected in the second part of the analytical tool set by AFI.

Parameters	Definitions/values		
Maintenance nominal life (L_m)	Expressed in years		
Time from last maintenance intervention (T_u)	Expressed in years		
Investigation tests knowledge level (L_k)	Exhaustive (1.0)	Moderate (0.9)	Base (0.8)
Inspection date (T)	Expressed in year		
Previous inspection date (T_0)	Expressed in year		

4.2.3 The results offered by the analytical procedure set by AFI

The data collected by the trained staff are the base parameters used for the last evaluation part of the analytical procedure.

A timing factor for the inspections (Ft) is obtained by considering the elapsed time from the last inspection and the nominal life of the maintenance intervention, according to (1).

$$Ft = \left[e^{\frac{(T-T_0)-T_a}{(T-T_0)}} \right]^{\left(\frac{1}{L_m}\right)} \qquad (1)$$

where T= inspection date, T_0= last previous inspection date, T_a=inspection periodicity and L_m= nominal life of the maintenance intervention.

A second result is the maintenance timing factor (Fm), a measure planned on specific elements, calculated considering the elapsed time from the last maintenance intervention and its nominal life, as reported in (2).

$$Fm = \left[e^{\frac{(T_u-L_m)}{(T_u)}} \right]^{\left(\frac{1}{L_m}\right)} \qquad (2)$$

where T_u= elapsed time from the last maintenance intervention and L_m= nominal life of the maintenance intervention.

The operating state (So) of the analyzed element is determined by considering the definitions associated to the first basic parameters (see Table 1) and the above described factors Ft and Fm, as indicated in (3).

$$So = \left[\frac{a + (b \cdot c) + d}{3} \right] \cdot Lc \cdot \frac{\min(L_d)}{F_t \cdot F_m} \qquad (3)$$

where a= general state of conservation, b= new cracks, c= previous cracks, d= falling effects, Lc= investigation tests knowledge level; Ft= timing inspection factor and Fm= maintenance timing factor.

Based on the resulting operating state, five classes of alert levels can be determined:

- Alert level 1, when So > 75%,
- Alert level 2, when 45% < So ≤ 75%
- Alert level 3, when 20% < So ≤ 45%
- Alert level 4, when 10% < So < 20%
- Alert level 5, when So < 10%.

For each alert level, the evaluation of the time interval between the inspections is also rearranged: with the alert level 1, 2 and 3, the scheduled inspection period is maintained, whilst with the alert level 4 and 5 it decreases respectively as 0.75*Ta and 0.5*Ta (where Ta is the periodicity of the inspection).

5 DISCUSSION

The evaluation of the alert risk for specific architectural elements set by AFI represents an innovative tool specifically calibrated for specific damage typologies, mainly connected to decay process of the building and decorative elements like statues and moldings. After the accident occurred during 2017 in Florence, where the falling of a piece of stone from a corbel in Santa Croce Basilica killed a tourist, the analytical tool received the attention of the Italian Ministry for Cultural Heritage, showing the interest of the institution for topics related to safety aspects and risks mitigation.

The research team of Politecnico di Milano is working on some simulations on possible scenarios involving the describe procedure into different level of complexity. Even if the research is at an early phase, some observations can be drawn out.

The parameters are clearly oriented towards the immediate identification of the cracks, leaving limited possibility to provide a complete description of other damage types that could be present on the stone materials. Considering the specific case of the Milan Cathedral, knowing the characteristics of the Candoglia Marble, subjected to accelerated worsening process in contact with the polluted atmosphere of the city, more attention to those aspects could be introduced by additional sub-parameters that could refine the definition of the general state of conservation.

From a practical point of view, the definition of the areas, with the elements to be analyzed, constitutes another hard exercise. Due to the monumental dimension of the building, the electronic tool cannot be applied to simple units (or macro-elements). For example, the common span of an external prospect, hosting a window, is so extended that the common form for recording the required parameters is not sufficient for collecting all the information of a single element. The solution is a very time-consuming subdivision of the main elements into more detailed parts. An alternative could be provided by the new advanced 3D model of the cathedral: a future support could be offered by digital devises allowing a real time navigation of the various parts of the building with the possibility to record the information for each damaged element, stone by stone.

6 CONCLUSIONS

The long experience matured by the Veneranda Fabbrica in monitoring the state of conservation of the Milan Cathedral offered an important source of data for calibrating the risk assessment practices. Close to reliable procedures, founded on specific skills belonging to the operators working for the Veneranda Fabbrica del Duomo, new safety demands for public visitors are addressing the research of new methodologies for evaluating the active risks connected to cortical damages. Failure of the stone units, composing the surface decorations (statues, moldings, sculptures), requires specific tools for classifying alterations and damages in order to set a virtuous monitoring program based on inspections.

The analytical tool described in this paper is a new procedure based on the methodological premises of condition assessment addressed to the identification of risks. The evaluation procedure is calibrated for providing a localization of building elements subjected to evident damages, indicating possible future failure risks. The proposed electronic sheet calculates different alert indexes in order to prevent active risks by modifying the dates of the planned inspections in relationship with the materials and the structures characteristics and the evolution of the damage detected during periodical controls. The research team of Politecnico di Milano, actually carrying out some simulations by different decays scenarios for the same architectural elements (like decorated windows and spires), observed some difficulties in managing the risk evaluation procedure for large monumental elements and some limits for a more detailed definition of specific decay pathologies respect to the clear observation of cracks. As future perspective, the introduction of a complete 3D digital model of the cathedral could provide new supports for recording the information during the periodical inspections and refine the quality of the data referred to the various materials conditions, requiring interpretation for preventing the effects of recognized active risks.

REFERENCES

Achille, C., Bocciarelli, M., Canali, F., Coronelli, D., Della Torre, S. & Fassi F. 2018. The Duomo of Milan: some recent interventions on the main spire and on the tiburio vaults. In G. Milani, A. Taliercio & S. Garrity (eds), *10th International Masonry conference, Milan,* July 9-11, 2018: 2073–2086.

Aste, N., Adhikari, R.S., Buzzetti, M., Della Torre, S., Del Pero, C., Huerto Cardenas, H.E. & Leonforte, F. 2019. Microclimatic monitoring of the Duomo (Milan Cathedral): Risks-based analysis for the conservation of its cultural heritage. *Building and Environment,* 148: 240–257.

Gentile, C & Canali, F. 2018. Continuous monitoring the Cathedral of Milan: documentary and preliminary investigations. In G. Milani, A. Taliercio & S. Garrity (eds), *10th International Masonry conference, Milan,* July 9-11, 2018: 2061–2072.

Canali, F., Cigada, A., Lucaccioni, C. & Toniolo, L. 2018. Enhancing a gothic cathedral with 3rd millennium technologies: the Duomo di Milano. In G. Milani, A. Taliercio and S. Garrity (eds), *10th International Masonry conference, Milan, July 9-11, 2018:* 2103–2113.

Opera della Primaziale pisana, 2012. *Atti del convegno Cattedrali europee: conservazione programmata, 18-19 maggio 2012.* Pisa: Opera della Primaziale pisana.

Camuffo, D. 1997. Perspectives on risks to architectural heritage. In N. S. Baer & R. Snethlage (eds), *Saving our architectural heritage: the conservation of historic stone structures:* 63–92. New York: John Willey & Sons Ltd.

Clark, K. 2014. Values-Based Heritage Management and the Heritage Lottery Fund. *UK. APT Bulletin: The Journal of Preservation Technology,* Special issue on values-based preservation, 45(2/3): 65–71.

CEN/TC 346 – Conservation of Cultural Heritage. 2012. *Conservation of cultural property – Condition survey of immovable heritage,* EN 16096: 2012.

CEN/TC 346 – Conservation of Cultural Heritage, 2017. *Conservation of cultural heritage- Conservation process – Decision making, planning and implementation,* EN 16857: 2017.

CIB Commission W023 – Wall Structures, 2010. *Guide for the structural rehabilitation of heritage buildings.*

Regione Lombardia, 2003. *La conservazione programmata del patrimonio storico architettonico. Linee guida per il piano di manutenzione e consuntivo scientifico.* Milano: Regione Lombardia. Direzione generale culture, Guerini & associati.

Della Torre, S. 2010. Critical reflection document on the draft European Standard CEN/TC 346 WI 346013 Conservation of cultural property – Condition survey of immovable heritage [unpublished discussion document].

Della Torre, S. & Cantini, L. 2018. Damage control, preservation procedures and durability studies: an investigation approach through the Milan cathedral archives. In G. Milani, A. Taliercio and S. Garrity (eds), *10th International Masonry conference, Milan, Italy July 9-11, 2018:* 2114–2123.

Della Torre, S. 2018. The management process for built cultural heritage: preventive systems and decision making. In K. Van Balen,.and A. Vandesande, (eds), *Innovative built heritage models:* 13–20. London: Taylor and Francis Group.

Forleo, D. (ed) 2017. EPICO – European protocol in preventive conservation, Methods for conservation assessment of collections in historic houses, *Cronache 7,* Genova: SAGEP Editori.

eu-chic.eu., 2009. Eu-chic project official website. [on line] Available at: http://www.eu-chic.eu/[Accessed 10 Oct. 2016].

Feilden, B.M. 1982. *Conservation of Historic Buildings:* p. 3 London: Butterworth scientific.

Fermo, P., Goidanich, S., Comite, V., Toniolo, L. & Gulotta, D. 2018. Study and Characterization of Environmental Deposition on Marble and Surrogate Substrates at a Monumental Heritage Site, *Geosciences,* 8: 349.

Fry, C., Xavier-Rowe, A., Halahan, F. & Dinsmore, J. 2007. What's causing the damage! The use of a combined solution-based risk assessment and condition audit. In Padfield & K. Borchersen (eds), *Museum Microclimates:* 107–114. Denmark: T. National Museum of Denmark.

ICOMOS, 2000. *Heritage at Risk. ICOMOS World Report 2000 on Monuments and Sites in Danger.*

ICOMOS – ISCARSAH International Scientific Committee for analysis and restoration of structures of architectural heritage, 2003. *Recommendations for the analysis, conservation and structural restoration of architectural heritage.*

ISO International Organization for Standardization, 2010. *Bases for design of structures – Assessment of existing structures,* ISO 13822:2010.

Kioussi, A., Karoglou, M., Bakolas, A., Labropoulos, K., & Moropoulou, A. 2013. Documentation protocols to generate risk indicators regarding degradation processes for cultural heritage risk evaluation. In *International Archives of the Photogrammetry, Remote Sensing and Spatial Information Sciences, vol. XL-5/W2, 2013, XXIV International CIPA Symposium*, 2-6 September 2013, *Strasbourg, France*: 379–384.

Michalski, S. 1990. An overall framework for preventive conservation and remedial conservation. In *ICOM Committee for Conservation, 9th triennial meeting, Dresden, German Democratic Republic*, 26-31 August 1990, *preprints*: 589–591.

Naldini, S. Heinemann, H. & van Hees, R. 2018. In K. Van Balen & A. Vandesande, *Innovative Built Heritage Models*:117–124. London: Taylor and Francis Group.

Stovel, H. 1998. *Risk preparedness: a management manual for world cultural heritage*. Rome, ICCROM.

Van Balen, K. & Verstrynge, E. (eds) 2016. *Structural Analysis of Historical Constructions – Anamnesis, diagnosis, therapy, controls*. London: Taylor & Francis Group.

Waller, R. 1995. Risk management applied to preventive conservation. In C. L. Rose, C. A. Hawks & H. H. Genoways (eds), *Storage of natural history collections: A preventive conservation approach*: 21–27. New York: Society for the preservation.

Preventive Conservation - From Climate and Damage Monitoring to a Systemic and Integrated Approach – Vandesande, Verstrynge & Van Balen (eds)
© 2020 Taylor & Francis Group, London, ISBN 978-0-367-43548-6

The role of the university in maintaining vernacular heritage buildings in the southern region of Ecuador

G. García, A. Tenze & C. Achig
World Heritage City Preservation Management Project, Faculty of Architecture and Urbanism, University of Cuenca, Cuenca, Ecuador

ABSTRACT: Over the last decade, the University of Cuenca (UC), located in southern Ecuador, through its Faculty of Architecture and its City Preservation Management (CPM) project, has progressively become a key player in addressing the complex task of conserving architectural cultural heritage. In particular, UC has directly contributed to broadening the scope of conservation practices by encouraging the protection of modest examples of vernacular earth-based architecture. This article reflects on innovative conservation strategies for this type of UC-led architecture, where the greatest strength has been the adoption of a participatory approach in different urban and rural contexts considered as living labs for research. Besides, this contribution shows the potential role that the academic actor can play as a platform for the exchange of scientific and empirical knowledge at a local and international territorial scale.

1 INTRODUCTION

1.1 Maintenance of vernacular architecture

Vernacular architecture has recently been considered as part of the products of cultural heritage. During the 20th century, the archaeological, historical and aesthetic scope of its study was broadened to include social and technical aspects. At that time, the aesthetic aspects of vernacular architecture were the source of inspiration for well-known architects such as Adolf Loos, Frank Lloyd Wright, Le Corbusier, Alvar Aalto, among others. Progressively, from the social point of view, experts such as P. Oliver and A. Rapoport explored a new perspective on their study, emphasizing on its balanced relationship between human beings and their natural environment. It arises as part of a trial-and-error process that is transmitted from generation to generation, permanently adapted and transformed. In terms of P. Oliver (2003) vernacular architecture is "architecture built by and for people". Through his study, one could reveal the social dynamics of each community, including their preferences or desires, etc. Consequently, unlike other types of architecture, vernacular architecture is considered a vast repository of identity. According to ICOMOS (1999), vernacular architecture: "is the traditional and natural way in which communities accommodate themselves. It is a continuous process that includes the necessary changes and continuous adaptation in response to social and environmental constraints".

Within vernacular architecture, earthen architecture corresponds to one of the most widely used building systems in the world. Until 2011, about 50% of all earthen architecture was in the southern parts of America, Africa, Asia, and Europe. According to CRAterre (2012), about 150 of the world heritage sites listed by UNESCO correspond to earthen architecture, and most of them (26%) are located in Latin America. For the past fifty years, the study of earthen architecture has gained prominence in academic debates. From the point of view of its physical characteristics, important advances have been consolidated and centers and networks of very respectable specialists focused on its improvement have been installed all over the world. As a result of this type of study, earthen architecture has been permanently referred to as a source of knowledge from the past, highlighting the lessons learned to face contemporary problems and even beyond them. However, for the time being, earthen architecture has progressively entered into decline and disuse, both in urban and rural areas.

In Ecuador, earthen architecture has ancient roots (pre-Inca period), which were adopted during the Colonial period and the Republican period and still persists, especially in the Andes Region. In fact, the second province in this territory with the largest number of adobe buildings is Azuay (García et.al. 2016). In this province, vernacular earthen architecture has an important presence not only in rural areas but also in urban ones, such as the Historic City Centre of Cuenca, which was listed as a World Heritage Site in 1999. Despite the relevance of vernacular earthen architecture in Azuay, it can be considered one of the most threated architectural heritage. In fact, until the first decade of the twentieth-first century, efforts from the public sector prioritized the conservation of monumental structures that represented less than 1% of total heritage buildings listed. Currently, in a socio-economic context, guided by economic pressures and changes on cultural preferences introduced mainly due to migration phenomena, the query was, how to favor the conservation of vernacular earthen architecture?

1.2 The role of the University

The Constitution of the Republic of Ecuador (AN 2008) states that the Higher Education System must provide academic and professional training with scientific and human vision. Therefore, any Ecuadorian university has three substantial functions: education, research, and articulation with the needs of society (AN 2010). By this regulation, Universities are advocated to coordinate efforts among them and with civil society, observing principles such as social justice, equity, participation, and co-responsibility (AN 2010). However, there are few examples of Ecuadorian Universities on which these three substantial functions and principles have been integrated simultaneously in their initiatives.

Within the architectural heritage arena, one exceptional case corresponds to the University of Cuenca (UC). The UC is a public academic institution whose mission is to provide high-quality education to enhance the well-being of communities, respecting their cultural and natural environments. It was founded in 1867, in the city that shares its name. It is considered the principal university of Azuay Province, located in the south of the Ecuadorian country. It offers a wide variety of degrees and international master's programs, all supported by high-quality and innovative, interdisciplinary research.

Since 2007, the conservation of architectural heritage has been led by the Faculty of Architecture and Urbanism. Within this academic space raised the Vlir 'World Heritage City preservation management' project, as a result of an inter-institutional agreement between the Belgian KU Leuven and the UC. That inter-institutional agreement lasted one decade and contributed to strengthening scientific research and academic training in closed collaboration with the

Raymond Lemaire International Centre for Conservation, Leuven.

Thanks to that impulse, the UC has progressively become a key player on facing the ambitious task of conserving vernacular earthen architecture located in urban and rural contexts, in the South of the Andes region of Ecuador. Initially, based on a preventive conservation approach, the CPM project focused on (i) documenting and monitoring activities which resulted in a GIS-based monitoring system in Cuenca, (ii) planning development of historic cities based on the Historic Urban Landscape (HUL) approach, (iii) analyzing heritage as a resource of development, which resulted in a participatory methodology for decision-making process. The resulted from qualitative, quantitative and participatory data allowed to identify constraints and needs and gave the reasons for the continuity of the CPM project with new research lines that are currently in progress (Table 1).

1.3 Participation in the cultural heritage field

Although social participation has been increasingly accepted in cultural heritage discourse over the past decades, the practical implementation of the participatory approach often seems ambitious and is not easy to apply. Unfortunately, in several countries - including Ecuador - the participatory approach has been affected by growing skepticism. According to Villasante (Alberich et al 2015), this due to repeated misleading consultative practices from the political side, aggravated by the low interest of social actors in being part of a genuine participatory planning process.

In the last four years, fed by the contributions of the social field, the UC has progressively enriched a theoretical framework and has consolidated practical experiences to deal especially with the maintenance of vernacular earth architecture. The Maintenance Campaigns correspond to one of the experiences in which different actors adopted preventive practices. This implies small but significant activities that differ from restoration activities, e.g. repair of roofs, external walls, and water evacuation systems.

The Maintenance Campaigns, observe the ICOMOS preventive conservation methodology

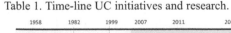

Table 1. Time-line UC initiatives and research.

Ecuador- South America//   Province of Azuay// Cuenca(top)-Susudel(bottom)

Figure 1. Location of UC living-labs.

(2003) which states at least four main phases: 1. Anamnesis, 2. Diagnosis, 3. Therapy and 4. Control. To their operationalization, the maintenance campaign model was inspired by the system of a community organization that is still alive in some rural areas in Latin America, known as 'Minga' (Cardoso et al2019). Minga is a pre-Hispanic practice that refers to collective free work in order to achieve common benefits. In this case, it articulated different actors, initially: the academic (who until now has assumed the role of leader and coordinator), the public institutions (legally responsible for cultural heritage conservation), and community members (owners). Those actors were organized in different teams, mixing different technical and practical skills, in a human structure which includes for each group: Professors, students, soldiers, community members, and one master mason. Each group had its own name (proposed by the community members) with the intention of strengthening the sense of belonging and creating a spirit of commitment among the teams.

The first maintenance campaign took place in a rural area of the province of Azuay, called Susudel, in December-2011, and later was applied in two traditional urban neighborhoods of the city of Cuenca, in the same province (Figure 1). Therefore, through this experience was boosted connections among different actors, community and academy members, and public institutions in favor of heritage preventive conservation based on traditional knowledge. Since then, the maintenance campaigns have been refined, towards a more participatory approach.

2 METHODOLOGY

Taking the four maintenance campaigns as an initiative easier to monitor (2011-2018), this article presents a comparative analysis of the process, people involved and results obtained on each case. Using quantitative and qualitative data, some of the main changes towards a more participatory approach were identified.

2.1 Maintenance campaigns in the rural area of Susudel

Susudel is a rural area located in the province of Azuay in the South of Ecuador. Its economy is based on agriculture (self-consumption) and brick production. This last is considered the main source of income (43%). Brick and adobe production are considered traditional activities for its inhabitants who have taken advantage of the quality of the land where they live. Susudel has about 1188 inhabitants, the majority (53%) are female and 11% of its population is more than 65 years old (Izquierdo & Orellana 2013). In contrast with the colonial settlements in urban areas in the country, the settlement of Susudel resulted from a colonial production mode extended through the Andean region, called *Latifundia*. In those Latifundia, indigenous worked in exchange for food and a place to live. This system lasted until the Agrarian Reform in 1964. After that, the workers took possession of the land and they occupied the proximity of the ancient access called *'chaquiñán'*.

Since then, Susudel has grown on the edge of the main access road, rather than around a square. Buildings are mainly devoted to housing (84%), with a low density of 14.32 inhabitants/ha (Moscoso 2009). In 2011, recognizing the knowledge available on traditional building systems and the high level of community organization of its inhabitants, UC decided to promote the first maintenance campaign, in collaboration with local institutions and community members. It was applied to conserve a set of vernacular architectural heritage, used as dwellings, in the street of the Possessions and in 2013 was implemented the same model of join work for the conservation of a public good, the cemetery of Susudel.

The maintenance campaigns were preceded by meetings with members of the community of Susudel. In those UC-led meetings, the criteria for the selection of the buildings were explained, such as the heritage value of the property, the socio-economic condition of the owners and their willingness to participate in the process. Besides, considerable efforts were made to involve public institutions in the process. During the first campaign, a significant amount of time - more than three months - was required for preparing the work, which was executed in five days. Unlike during the second campaign, less time was invested mainly due to the collaboration of the parish authority. This last, met with neighboring communities such as Sanglia, San Gerónimo, Susudel Centro, Susudel Nuevo, Raricucho, among others, to give a full day of work to the execution of the minga during the maintenance campaign of their common cemetery. In both campaigns, some people participated as workers during the intervention, while others provided food for about 150 people (Cardoso et al. 2019).

2.2 Maintenance campaign in the urban area of Cuenca.

The collaborative work experienced in the rural area of Susudel moved to the urban area of

Cuenca. This is the third most populated city in Ecuador (506,000 inhabitants), located in the Andean region. Its historical area was declared a World Heritage Site by UNESCO in December 1999. The last inventory of heritage buildings (2010) recorded almost 3,154 heritage buildings. The architecture of Cuenca is a remarkable example of urbanism and architecture that integrated the diverse local and European influences, many of which date back to the 18th century, and which were "modernized" in the economic prosperity of the 19th and early 20th centuries. Most of this architecture illustrates the traditional techniques of earthen architecture, which include decorative elements on the façade, interior spaces with decorated surfaces, details such as moldings, stuccoes, walls, and ceiling decoration. However, this architecture is extremely fragile due to the low resistance of its materials. According to the Preventive Conservation Diagnosis Plan (DAHP 2011), 8% of all heritage buildings require urgent intervention and 18% medium-term intervention.

In the urban context, UC selected two sets of heritage buildings located within the historic area of Cuenca, to apply the maintenance campaign model imported from rural context. The first set of buildings (24) corresponded to one block of the neighborhood of San Roque (2014), while the second set of buildings (15) were located around the corridor of the neighborhood El Vergel, called Calle Las Herrerías (2018). Both areas shared a common physical organization that has been partially preserved since 1920-1950. Besides, the physical characteristics of the heritage buildings were similar to those of the heritage buildings of Susudel, that is, modest examples of vernacular earthen architecture, in a considerable state of deterioration. The criteria of selection of heritage buildings were the same originally adopted in Susudel in 2011.

The planning process in the urban area demanded more time than in the rural area. It was almost a year of preparatory technical work and community meetings for the organization between community and institutions. In this process of organization, it was very helpful to share - in a video - the experience developed with the community of Susudel. This provided a general idea about the objective of the process, the actors involved and the achievements. Although both the characteristics of the heritage and the activities developed were similar to those of the rural area, however, the urban interventions lasted about two months in each case.

2.3 Comparative analysis.

In all cases, the implementation of the maintenance campaign model favored the collaboration and involvement of different actors. Excluding the private companies of San Roque, all the campaigns shared almost the same actors. Table 2 clearly shows the greater participation of the academic actor represented by students, professors, and researchers. The Academic actor is precisely the actor that promoted and coordinated these experiences. In the rural campaigns, institutional actors such as the Municipality of Oña and the Government of Azuay, through the support of the army as a labor force, shared the protagonism with the UC. In the Susudel Campaign (2011) the National Institute of Cultural Heritage (INPC) intervened while in the cemetery (2013) the parish authority acquired a fundamental role in calling communities to work in minga.

Concerning the maintenance campaigns in the urban area, the participation of the local Municipalities (financing) and the army (labor) was fundamental for the implementation. It is worth noting that in the urban context, there is increasing participation of organizations that, through a minga, support maintenance campaigns in the city of Cuenca. For example, there are public companies such as the Electric Company, and private companies such as Drinking Water Company, among others; NGOs such as the Salesiana PACES Foundation, which provided support through internships for students, who are training in trades related to construction and have even incorporated other local universities.

Related to the physical conservation of heritage buildings, the availability of knowledge (know-how) about traditional constructive systems established an important difference in the results obtained in each context. In the rural context, almost 100% of the activities implemented followed traditional techniques, while in the urban context, the lack of knowledge, traditional materials, time pressures, and

Table 2. Actors involved in maintenance campaigns.

Actors		RURAL		URBAN	
		2011	2013	2014	2018
Academic Actor	Students	14	6	39	55
	Professors and Researchers	5	4	10	16
P.Public institutions		3	3	8	14
Asociations		x	x	1	3
Private Company		x	x	5	x
Community	Intervened buildings (beneficiary families)	49	1	22	20

availability of new materials, influenced the expected results. It means in Susudel it was possible to preserve the integrity of built heritage, more than in Cuenca.

The comparative analysis between Susudel and Cuenca on the participation of landowners as a labor force during the intervention process shows scarce participation in the urban area. In fact, in urban areas, the financial contribution of community members was required to resolve the scarcity of knowledge concerning traditional construction systems. Using the example of Susudel (2009) and Cuenca (2014) tables 3 and 4 show respectively the number of participants during the intervention process, classified by type of work, external (E) or local (L), where local work refers to neighborhood residents. The comparison shows that the number of qualified masons required in an urban area is twice the number required in the rural context to do half the work.

Table 3. Susudel´s participants by type of labour.

PARTICIPANTS	NUMBER	LABOUR TYPE
Students	17	E
Professor	5	E
Soldiers	43	E
Qualified manson	10	L
Community members	60	L
Students – community	36	L
Professionals	5	E
TOTAL	176	

Table 4. San Roque´s participants by type of labour.

PARTICIPANTS	NUMBER	LABOUR TYPE
Students	37	E
Professor	5	E
Soldiers	30	E
Qualified Manson	23	E
Community members	0	L
TOTAL	95	

Nevertheless, during the post-campaign evaluation process, a greater level of awareness of the importance of traditional techniques was identified among community members in both contexts:

"The university project is important because it has changed our way of thinking. We had the idea of leaving the adobe and the tiles to prefer new materials, but the university has revived the idea of making our houses in the same way as before. Now, we say we are going to restore, we are going to promote... since then, the interest to preserve the traditional techniques

is increasing, even in the new buildings" (Susudel: 2012).

"The campaign has raised awareness of maintaining and preserving heritage buildings", "It was a lesson in solidarity with mutual benefits. Residents have improved their living conditions and the city as a whole has been beautified". (San Roque residents: 2013).

Exceptionally, participatory methodologies were included in the Herrerías campaign, not only surveys and interviews were carried out. Interactive and participatory activities such as guided visits to homes walk around the neighborhood, SWOT analysis, mapping of actor-networks, and even role-playing dramatized by the neighbors themselves, creating positive and negative scenarios for future work. Through this participatory process of action research, UC sought to improve each phase of the preventive conservation process and carry out a shared work based on the formulation of the project among the different stakeholders, the community being one of the most important to take into account. In Las Herrerías, maintenance campaign rescue heritage, but even more, this process of improvement benefited the activation of the social networks of the neighborhood with a more prominent role.

3 RESULTS

3.1 Permanent refinement of participatory maintenance of vernacular heritage

By investing time in building a relationship with the territory and inhabitants, progressively UC has adopted a very active listening, but fundamentally UC has boosted deliberation among the actors involved as part of the dialogue for improving the technic initiatives. It goes in line with the awareness of participation as a reaction against the unbalanced power observed among groups to influence the decision-making process. In line with this, the UC has contributed to reflect on the struggles between the so-called 'experts' (academics, professionals, institutions) and 'non-experts' (citizens) (García 2018).

This has implied for the academic actor (UC) becoming open to including the empirical experts' perspectives. In fact, instead of implementing their desired projects, be open to creating the possibility to inquire about their ideas and debate together about them. Progressively, community members have been included in the early phases of the ICOMOS process (table 5). In the end, those experiences have strengthened local capacities through a two-fold learning process –scientific and empirical- moving from initiatives built based on an interdisciplinary

Table 5 . Actor's participation in campaign phases.

Maintenance Camping (location and year)		Phase 1 Analysis	Phase 2 Diagnosis	Phase 3 Therapy	Phase 4 Control
RURAL	2011	A	A	A/C/PI	A
	2013	A	A	A/C/PI	A
URBAN	2014	A	A	A/C/PI/Aso/PC	A
	2018	A	A/C	A/C/PI/Aso/PC	A

*Actors: Public Institutions (PI), Academic actor (A), Associations (Aso), Private Companies (PC), Community members (C)

perspective towards the co-creation, based on the exchange of these two types of knowledge.

Moreover, considering participation only has a sense when its decisions are implemented, UC has drafted a strategy for sharing responsibility as a way to react against the populism and the paternalism that usually affects this kind of processes. In that strategy, the UC played a key role as a mediator actor among the rest. Progressively, all of the actors –including researchers- have transited toward the less frequent stage of participatory planning where citizens, as empirical experts, are also being considered to build proposals together and to define shared responsibilities.

3.2 Current and further perspectives

Those experiences have served to reveal that intangible traditions of collective management in local communities, such as *Minga* and *Ainy*, might constitute an untapped resource for creating better governance approaches that contribute to the sustainable management of vernacular architecture and thus living conditions. Capitalizing this acknowledgment has served to articulate the UC experience in two projects with international collaboration concerning to cultural heritage management: (i) the project entitled "Innovative governance systems for built cultural heritage, based on traditional Andean organizational principles in Ecuador" (TEAM VlirUOS), and (ii) the International network for Leveraging Successful Cultural heritage-led Innovations and Diplomacy through capacity building and awareness-raising (ILUCIDARE).

In the first case, the project resulted winner in the last international call for research projects Vlir-TEAM 2018 and will run from January 2019 to December 2022 (4 years). It seeks to deepen the understanding of and activate the traditional Andean knowledge on organizational principles/collective management in the South-Eastern Ecuadorian Highlands. It explicitly opts to combine traditional Ecuadorian knowledge with the best available science, based on an approach that emphasizes interdisciplinary as well as the participation and capacity building of relevant stakeholders in different representative Test Beds. This co-creation approach towards innovative governance guidelines ensures that the developed results are user-oriented and hold a long-term vision and pathways for policymakers, public, and private actors.

Invaluable knowledge has been gained by the various actors involved in past research and case studies. They have demonstrated the ability to get scientific results and social impact. Besides, those experiences have revealed the need for involvement of more disciplines to strengthen the governance component to develop more sustainable self-governed approaches. In line with this, the TEAM project aim to join university actors together with other related stakeholders to become: 1. More susceptible to participatory and inter-disciplinary projects, 2. Innovators with new capacities and attitudes, 3. trainers sharing evidence-based practices.

On the other hand, ILUCIDARE is a three-year project that aims to promote and leverage Cultural Heritage (CH)-led innovation and diplomacy through the creation and activation of an international community of CH practitioners in Europe and beyond, while strongly contributing to the overall objectives of the communication towards an EU strategy for international cultural relations and EU international cooperation in research and innovation. ILUCIDARE refers to both "elucidare", aiming to provide a common definition of CH-led innovation and diplomacy, and "lucidare", aiming to raise awareness and engage people in a new way through co-creation approaches to favor paradigm shifts in CH diplomacy.

In line with the project previously described, ILUCIDARE's approach and methodology emphasizes interdisciplinary and participatory approaches towards CH to directly involve and empower stakeholders (government, businesses, research/education, and civic actors), but on this particular case, through the link of local networks and actors with strong international networks of the cultural heritage field, which include organizations such as Europa Nostra, the leading umbrella organization in the CH field in Europe, and officially recognized NGO partner of UNESCO; the World Monument Fund, Kosovo Foundation for Cultural Heritage without Borders (CHwB), prominent research and education institutions such as the KU Leuven and the International Cultural Centre (ICC); KERN European Affairs

(KEA) which has built an unparalleled community of cultural actors though social media and outposts in Europe and China and has extensive experience in developing and managing international networks; Interuniversitair Micro-Electronica Centrum (IMEC) is Belgian not-for-profit research and technology organization, world-leading research and innovation hub in nanoelectronics, digital technologies, living labs, and user research.

The project also counts on the support of the African World Heritage Fund and World Heritage Institute of Training and Research for the Asia and the Pacific Region (WHITRAP) to further support the capacity building outcome. This link local-international will enable the exchanges of best practices, knowledge transfer, skills development, and cross-fertilization within a global network through extensive use of digital engagement strategies and tools as well as participatory activities. The main legacy of ILUCIDARE will consist of an international community for leveraging successful CH-led innovations and diplomacy through engagement and capacity building actions, along with the project tangible outcomes.

Both projects on which the UC is currently working, target the main goals of Ecuador's country strategy, namely becoming 'a country of knowledge' whereby the focus is both on scientific and ancestral knowledge that serves in the conservation of vernacular architecture to two central themes (i) capacity building and training of individuals and (ii) developing excellence within higher education institutions. Here, UC act as a sounding board, interplaying in project meetings with a diversity of actors and disseminating knowledge through local events in Ecuador and different international events. It means the role of UC has become crucial in building the platform for exchange between international and local actors of rural, peripheral and urban areas. In that sense, UC has expanded previous areas considered as living-labs, such as Cuenca and Susudel to include Saraguro, Nabón, respectively (Test Beds) in the southern region of Ecuador.

4 CONCLUSIONS

The variety of initiatives promoted by the UC in favor of the conservation of vernacular architecture has resulted in various lessons learned. One of them was the need to develop joint initiatives between various actors to optimize scarce local resources (human, financial and temporal). However, at the same time, it was one of the biggest challenges. On the one hand, there is no linear process to identify the purposes of sharing to mobilize actors, and even more challenging is to sustain joint work in time.

In this regard, UC has learned that its potential is not enough. Although more and more actors might be integrated into the processes, such as maintenance campaigns, not everyone gets involved in what is considered real participation. For instance, the fact that political events influenced public institutions the decision of being part of these processes –or not-. It made us think that it is still necessary to internalize the concept of shared responsibility, not only from an academic reflection but also better from a reflection-action, which proposes go beyond a project or initiative, towards a sustained process that should be based on the synergy of all actors instead of a new vertical structure with a different actor at the top.

Through the initiatives in progress, UC aims a new scenario for the heritage conservation based on shared protagonism. It implies not only broadening the methodological framework, but also articulating efforts from the first stages of any process, and moving from a short-term to a long-term approach. UC recalls the idea of participation as something different from a meeting of the parties or collaboration at specific moments. Participation includes in itself the ancestral principles of solidarity, commitment, co-responsibility, which is combined with the natural creation of networks that one can be observed in daily life. This paradigm guides the different initiatives of the UC, with their particularities, but always articulated in the recognition of the network of specific actors in each circumstance and territory.

UC has given important steps towards this participation paradigm. Initially, by promoting the articulation of diverse local actors, and then by being a platform for dialogue that extends this dialogue even to international actors. Initiatives such as the maintenance campaigns, guided from a concrete action, favor the dialogue. Progressively, these initiatives have become in spaces for listening, where exchange transcends into the configuration of cooperation networks. All this is facilitated by the use of methodological devices that motivate joint reflection and self-reflection. In this direction, UC continues working, but this task implies that the UC takes on new challenges, some of which are:

1/Being open to territorial networks, from a subject-subject approach, for example, by considering the members of the community as "empirical experts" instead of "non-experts" or by not considering them only as objects of research, but as protagonists of the action; 2/Encouraging the creation of initiatory or (pro)motor groups that transcend the academic space, considering the integration of interested (non-representative) people with a territorial base; 3/Activating follow-up commissions with the most representative actors, to link each experience or process to the collaboration of those who can move diverse resources, organizing with these meetings systematic work overtime; 4/ Developing methodical systems that take into account those instruments or devices more creative, to articulate and consolidate not only the joint work but also to stimulate synergies that promote other more similar experiences, even in other areas of life in the territories.

With these contributions, we hope to have a positive impact on society, not only at the territorial and community level but also at the academic level, developing in each initiative models of co-creation that articulate actors and networks of actors that support the conservation and enhancement of the vernacular heritage and beyond.

REFERENCES

Alberich et al., Metodologías participativas: sociopraxis para la creatividad social. RedCIMAS. Madrid: Dextra, 2015.

Asamblea Nacional del Ecuador, Constitución de la República del Ecuador. Ecuador, 2008.

Asamblea Nacional, "Ley Orgánica De Educación Superior Ecuador, Presidencia de la República del," Ley Orgánica Educ. Super. del Ecuador, pp. 1–63, 2010.

Cardoso, F., Rodas, C. & Achig, M., "CIUDAD-LABORATORIO: la enseñanza en la Universidad de Cuenca-Ecuador con las Campañas de Mantenimiento del Patrimonio". 19 SIACOT: Seminario Iberoamericano de Arquitectura y Construcción con Tierra. Oaxaca, México. 2019.

CRATERRE, Gandreau, D; Delboy, L; Joffroy, World Heritage Earthen Architecture Programme. 2012.

DAHP "Diagnosis of Preventive Conservation Plan". Municipality of Cuenca. 2011.

G. García, J. Amaya, & S. Ordoñez, "Desafios de los procesos de producción y construcción en adobe en América Latina, retos y oportunidades," in 16th SIACOT 2016, pp. 36–47.

G. García, "The Activation Process of Cultural Heritage as a driver of Development," Doctoral dissertation KU Leuven, 2018.

ICOMOS, "Charter on the Built Vernacular Heritage," Mexico, 1999.

ICOMOS, "Charter Principles for the analysis, conservation and Structural Restoration of Architectural Heritage," Zimbabwe, 2003.

P. Oliver, Enciclopedia de Arquitectura Vernacula. 2003.

Project proposal "Innovative governance systems for built cultural heritage, based on traditional Andean organization-al principles in Ecuador", 2018.

Project proposal "Innovative governance systems for built cultural heritage, based on traditional Andean organization-al principles in Ecuador", 2018.

Izquierdo, J & Orellana M. Estudio de factibilidad económica para la producción y comercialización del amaranto en la parroquia Susudel del cantón Oña. Thesis. Universidad Politécnica Salesiana. Cuenca; 2013.

Moscoso, S. Technical report to listed of Susudel as National Heritage Site. Vlir CPM research project. Cuenca; 2009.

Damage diagnosis and monitoring of case studies

MDCS - a system for damage identification and monitoring

Rob P.J. van Hees

*Faculty of Architecture and the Built Environment, Delft University of Technology, Delft, The Netherlands
R-Kwadraat MonumentenAdvies, Zoetermeer, The Netherlands*

Silvia Naldini

Faculty of Architecture and the Built Environment, Delft University of Technology, Delft, The Netherlands

ABSTRACT: Before starting any interventions related to the maintenance or restoration of a monument, a thorough investigation is necessary, aiming to assess damage to materials and structures and to obtain a well-supported diagnosis. This paper is based on a keynote lecture given in Leuven at the WTA-Precomos Conference on Preventive Conservation in April 2019; it discusses the importance of the use of so called damage atlases and presents the possibilities of using MDCS, the Monument Diagnosis and Conservation System, for diagnosis and monitoring. Originally created in 1995 within an EU project, MDCS has been developed into an online tool supported by the Cultural Heritage Agency of the Netherlands, TNO (Netherlands Organisation for Applied Scientific Research), and the Faculty of Architecture of Delft University of Technology.

MDCS includes Damage Atlases allowing a uniform identification of the damage found. They form the base for making hypotheses on damage mechanisms and for performing visual monitoring. Apart from the use by professionals, visual monitoring can also be performed by owners, which encourages their active participation in preventive conservation. The Damage Atlases are being used by several Provincial Monumentenwacht Organisations in The Netherlands to enhance uniformity and quality of Monumentenwacht activities on preventive conservation. The system is also used to support the Dutch Governmental Programme on Professionalism, aiming at improving restoration quality, through implementation in the guidelines of the Foundation ERM.

In the work of the WTA working group on damage monitoring, the option for visual monitoring will be proposed and used.

1 INTRODUCTION

Monitoring is the key to conservation. Well-structured, repeated inspections of a building allow to timely notice the occurrence or the increase of damage and to intervene accordingly, thus preventing further deterioration. In the Netherlands the Monumentenwacht organisation works since the 1970's following this assumption (Van Balen & Vandesande 2018). Monitoring can be used for predicting the extent up to which an intervention can be postponed, or to assess risk. It is also fundamental for evaluating the effectiveness of interventions, both at the level of the material and of the whole building. A basic requirement for effective communication in this context is the use of a common terminology related to degradation or damage. The importance of damage atlases with clear examples and definitions is therefore discussed in this contribution.

Monitoring means to assess selected conditions or properties and repeat the measurement or observation over time to be able to compare with previous results and draw conclusions. There are different types of monitoring, implying various skills and technologies: a clear definition of possibilities and limits of monitoring instruments and techniques is necessary to assign the correct value to the results. The Monument Diagnosis and Conservation System (mdcs.monumentenkennis.nl) supports visual investigations and monitoring. The application of the system in running programmes on heritage conservation in the Netherlands will be explained in this article. The contributions of MDCS to the ERM Guidelines for Quality in Restoration (ERM 2019) and to the WTA TC 7 Working Group Damage Monitoring (WTA) will be addressed.

2 THE IMPORTANCE OF A CORRECT DAMAGE DEFINITION

In the international context several damage atlases exist. A widely spread one is the ICOMOS *Glossary on stone deterioration patterns* (icomos.org). The *Damage atlas on damage patterns found in brick masonry* (Franke et al 1998) is the result of a EU framework programme.

Figure 1. Four times: degraded pointing. But, four different causes.

The importance of damage atlases lies above all in facilitating communication between involved people. Equally important is the fact that a correct definition of the observed damage or deterioration, also provides a first step towards a diagnosis. A sound diagnosis is necessary to intervene, without complications; in other words, to come to an intervention with compatible materials and techniques. Figure 1 shows four real life cases of damaged pointing and in all four cases the involved advisory bodies described the situation as '...pointing in a bad condition; it needs replacement'.

Even though this might seem correct, upon closer consideration, it is clear that in this way also a risk is involved. Just replacing the degraded pointing with what a contractor 'always uses' is exactly what often happens in restoration practice. However, there is a high risk that the same or even worse damage occurs. This is shown in Figure 2.

The new damage is clearly related to a lack of sound diagnosis and due to the contractor not being aware of the initial cause of damage.

A correct description according to a sound definition of the observed damage to the pointing would have allowed to propose a limited number of possible causes (hypotheses). This would have led to a diagnosis and recommendations (advice on material to be used, side measures) in order to avoid future damage.

Figure 2. Damage re-occurring after repointing: push-out of pointing (cf Figure 1c) and bending of masonry (as shown).

3 DEVELOPMENT OF DAMAGE ATLASES

The use of a standardised terminology can in the first place facilitate communication between involved parties, such as Monumentenwacht, restoration architect, contractor and eventually owner.

The damage atlas for brick masonry (Franke et al 1998), has been developed into a system of atlases with diagnostic possibilities, in a much broader field of heritage materials (including stone and also mortars and plasters,...). Supported by several projects from the EU Framework Programmes, the Monument Damage Diagnosis System, MDDS resulted (Van Hees et al 2009). Recently MDDS was transformed into a web-based application: the Monument Diagnosis & Conservation System, MDCS. This tool offers an interactive damage atlas and supports visual monitoring and evaluating the development of degradation over time.

In this context the system has broadened once more, now with the addition of the material concrete and with the introduction of structural damages, i.e. crack patterns and deformations.

An incorrect but recurrent way of identifying damage is to use terms like 'frost damage' which do not describe the form of decay but rather the (possible) cause. Starting from a correct damage type, one or more damage processes that could explain the damage type found, are suggested in MDCS. The damage process needs to be further assessed considering the circumstances under which damage has occurred.

The use of a damage atlas containing both terminology and possible causes leading to the damage type (or 'type of degradation'), offers an important step towards a compatible intervention. Such an approach is more likely to be a sustainable one.

4 MONITORING DEGRADATION

Visual monitoring forms an excellent support for studies and research on damage processes (Van Hees et al 2008). Some examples will be discussed.

An important damage phenomenon in historic buildings is related to salt crystallisation. For a sound diagnosis and intervention, it is necessary to understand the sources of moisture and salts and the environmental conditions under which the crystallisation damage develops. Further, it is important to take measurements over time; these generally include:

- the moisture and salt quantity and their distribution over the height and the depth of the construction
- the (indoor) climate conditions and their variations

These steps should be carried out, over a period of a year at least to understand the process and consequently propose an adequate intervention strategy. This in combination with other issues, such as:

Figure 3. Development of salt crystallisation damage over a period of 5 years (from MDCS).

- assessment of the salt types that are present
- types of materials concerned
- history and use of the building and interventions in the past

Visual monitoring by simply taking photos always of the same area and from the same position can help to understand the problem. Pictures taken by the owner of the monument or for example by an organisation like Monumentenwacht, during their yearly inspections, will give important additional information. This allows the researcher, who is usually only asked when damage reached an unacceptable level, to choose the relevant investigations for a better understanding of the process. Visual monitoring (Figure 3), used with the help of modern techniques for collecting information can be a great support to gain more insight in how the degradation process develops.

Figure 3 shows the development of salt crystallisation damage over a period of 5 years; here only the first and the 5th year situation are given for comparison. The pictures, combined with the damage definition and the judgement of the severity of the damage have been assessed using MDCS.

5 EASILY ACCESSIBLE USE BY BUILDING OR MONUMENT OWNERS

Potential use by monument owners is shown by a case in the North of the Netherlands. In the province of Groningen over the past years an increasing number of 'induced' earthquakes has occurred. The magnitude of the earthquakes, according to the Richter scale, seems rather low: until now, the strongest was 3.6 in the village of Huizinge. However, given the relatively limited depth at which the earthquakes take place (ca. 3 km), the peak ground acceleration is relatively high (around 0,080 – 0,11 g) and therefore the impact on buildings can be considerable, the more because they were never built to withstand earthquakes.

Figure 4. Use of a crack width meter.

In (Bal 2018) the necessity and importance of monitoring is underlined, but it also shows the difficulty of monitoring in a correct, meaningful and useable way. As the situation in Groningen is characterised by distrust between building owners and the authorities, a stronger involvement of the owners themselves in monitoring could be useful. In this way the professional research would be supported by additional, visual observation by people, directly involved, thus allowing to gather a large dataset of basically well-structured information.

Visual monitoring by owners of monuments offers an easily accessible possibility to capture a large dataset on damage development with simple tools like a camera (mobile phone) and a crack width meter, see Figure 4. In this way it is possible to assess the development of cracks and crack patterns, which are related to earthquakes. Even if a crack pattern cannot directly be assigned to earthquakes but for example to subsidence, it can become more manifest or develop to an unacceptable level due to the additional effect of the earthquakes.

6 BETTER UNDERSTANDING STRUCTURAL DAMAGE AND THE EFFECT OF INTERVENTIONS

Next to monitoring the development of a damage process, it is interesting to apply visual monitoring after interventions. In restoration practice, a case mostly closes even if it was based on investigations,

analyses, diagnosis and advice. Afterward, no more research or monitoring is performed, at least not until damage reoccurs a few years later and the owner starts to complain when it has reached an unacceptable level.

In cases of a successful intervention, nothing is even reported. This means that valuable knowledge, both on failure and on successful interventions gets lost.

It might be very profitable if a simple visual monitoring even on a yearly basis, would be performed and (anonymously) shared with the monument-community. Also here the use of MDCS offers this possibility, as is shown by a student's project performed at TU Delft and illustrated by Figures 5-7.

Here structural damage had been repaired and assessed in the following years. Figure 5 shows the situation before the intervention, which mainly consisted of filling-in the cracked zones with new masonry. Figure 6 gives a detail of the situation 6 years after intervention, which makes clear that damage, although less severe is re-appearing at a short distance from the original damage. It also shows (Figure 7) that the damage type (crack pattern) slightly differs from the original crack pattern; such observations might perhaps in future allow to fine-tune possible causes related to crack patterns or perhaps to combine patterns, although it is too early now for a definitive conclusion.

Figure 5, 6, 7. Monitoring crack patterns: Figure 5 before intervention, Figure 6, detail re-appeared crack, 6 years after intervention; Figure 7 comparison of crack patterns and severity of damage before and 6 years after intervention (MDCS).

7 USERS AND USER NUMBERS

MDCS can be considered a support tool for carrying out condition assessment and monitoring of monumental buildings. As indicated before, MDCS has been developed into an online tool supported by the Cultural Heritage Agency of the Netherlands, TNO (Netherlands Organisation for Applied Scientific Research) and the Faculty of Architecture of Delft University of Technology (monumentenkennis.nl & mdcs.monumentenkennis.nl). The system offers a method and the necessary background information to carry out inspections on masonry and concrete monumental buildings, make hypotheses on the causes of damage found, control the hypotheses and come to a diagnosis. Conservation techniques are presented to support interventions, when necessary. MDCS can also be used on the spot for the identification of the damage found.

MDCS has as an underlying purpose to improve the quality of restoration. As such it is intended to serve as an instrument for communication between architects, contractors and owners of historic buildings.

MDCS helps to clearly define and record damage types. This provides clear communication between parties in the construction industry: architects, contractors, inspectors, and insurance agencies. It is a first step towards determining causes and responsible procedures for intervention.

The user also learns to recognize materials. MDCS provides background information on degradation processes and maintenance and intervention techniques. Thus it is a source of knowledge both for professionals and students.

An important use of the MDCS lies in education, especially of restoration architects. Further, it has been introduced in the training of several provincial Monumentenwacht Organisations in the Netherlands (Monumentenwacht Moves). MDCS is available in Dutch and English.

On an international level, MDCS, being freely available and operating both in English and Dutch can be a useful instrument. It will for example, as a visual monitoring system, contribute to the work of WTA Technical Committee 7, working group on Damage Monitoring. Within the newly started CONSECH2 JPI project it will be implemented in the monitoring of specific aspects of historic concrete heritage buildings (Consech20); special attention will be paid to the involvement of volunteers in the care of historic concrete monuments.

The use of MDCS available through the website of Monument*en*Kennis is continuously monitored. Figure 8 and 9 show the number of users for 2018 and the first half of 2019. There is an increasing use of the system. Over 2018 in average there were 9 active daily users, over the first 5 months of 2019 the number was 20 active daily users.

In accordance with the new EU rules regarding privacy (General Data Protection Regulation – GDPR), it is not allowed to gather data on who exactly the users of MDCS are. Via direct involvement of the authors it

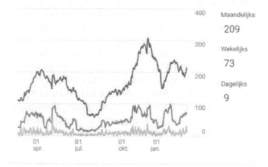

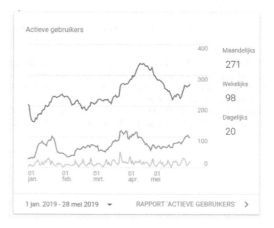

Figure 8, 9. Number of users for 2018 and first half of 2019.

is known that MDCS is used by Monumentenwacht in the Netherlands, by several academic education institutes and also by the Foundation for Conservation Quality - ERM, in their practical guidelines on interventions in historic buildings.

8 CONCLUSIONS

– MDCS is an online tool meant to support the diagnosis of damage and the conservation of monumental buildings; it is also a strong tool for performing visual monitoring.
– MDCS allows a uniform identification of damage based on damage atlases, which encourage a clear communication among users as this forms a necessary basis for monitoring.
– MDCS allows visual monitoring by professionals and by monument owners, the latter allowing to gather a crowd of information.
– The MDCS terminology is used by Monumentenwacht in the Netherlands, to achieve uniformity in the inspections, and is addressed by the ERM Condition Assessment Guidelines.

Next to the improvement of communication between parties involved in restoration processes,

MDCS can be used in education, especially of restoration architects. It has further been introduced in the training of several provincial Monumentenwacht Organisations in the Netherlands.

On an international level it will be used in the work of WTA Technical Committee 7, working group on Damage Monitoring. Within a recently started JPI CH project it will be implemented in the monitoring of specific aspects of historic concrete heritage buildings; this project will be focussing on the involvement of volunteers in the care of historic concrete monuments.

REFERENCES

Bal, I. 2018. *Myths and Fallacies in the Groningen Earthquake Problem*, inaugural speech, Groningen: Hanze University.

CONSECH20: CONSErvation of 20th century concrete Cultural Heritage in urban changing environments.

ERM foundation (Quality Monument Restoration): http://www.stichtingerm.nl/(retrieved Jan. 2019).

Franke L. et al. 1998, *Damage Atlas. Classification of Damage Patterns found in Brick Masonry*, Research Report No 8 Vol. 2, European Commission, Fraunhofer IRB Verlag.

ICOMOS: https://www.icomos.org/publications/monuments_and_sites/15/pdf/Monuments_and_Sites_15_ISCS_Glossary_Stone.pdf.

MDCS: https://mdcs.monumentenkennis.nl/.

MonumentenKennis: https://www.monumentenkennis.nl (retrieved Jan. 2019).

'Monumentenwacht Moves' - project Delft University of Technology, start at end 2018.

Van Balen, K. & Vandesande, A. (eds) 2018, *Innovative Built Heritage Models*, Leiden: CRC Press/Balkema.

Van Hees, R. et al. 2008. The use of MDDS in the visual assessment of masonry and stone structures. In Binda, L., di Prisco, M. & Felicetti, R. (eds), *On Site Assessment of Concrete, Masonry and Timber Structures, SACOMATIS, Proc. 1st Rilem Symposium, Varenna, 1-2 September 2008*: 651–660, Rilem Publications SARL.

Van Hees, R. et al. 2009. The development of MDDS-COMPASS. Compatibility of plasters with salt loaded substrates. *Construction and Building Materials*, Vol 23, No 5: 1719–1730.

WTA – Wissenschaftlich-Technische Arbeitsgemeinschaft für Bauwerkserhaltung und Denkmalpflege e.V. TC 7, Working Group Damage Monitoring.

Preventive Conservation - From Climate and Damage Monitoring to a Systemic and Integrated Approach – Vandesande, Verstrynge & Van Balen (eds)
© 2020 Taylor & Francis Group, London, ISBN 978-0-367-43548-6

Monitoring of water contents and temperatures of historical walls with interior insulation in Switzerland

Christoph Geyer, Barbara Wehle & Andreas Müller
University of Applied Sciences, Biel, Switzerland

ABSTRACT: Energetic refurbishments of walls in historical buildings with interior heat insulation show higher risks of damage caused by high moisture contents in the construction. Therefore, the robustness with regard to moisture protection of the construction is predicted by numerical programs, calculating the moisture content of the construction layers. Exterior walls of three historical buildings in Switzerland were refurbished with interior heat insulation. The insulation materials glass wool, stone wool and cellulose fiber were used. The moisture contents and temperatures of the material layers in these reconstructed walls were measured on-site during a time period of two years. The water contents of different layers of the reconstructed walls were calculated by using a numerical program for transient coupled heat and moisture transport. By comparison of the measurement values with the calculated ones the reliability of the prediction calculation was detected. The predicted values of the simulation program reproduce the measured values of moisture content and temperature in the construction layers within a small error margin. They could be used as a safety margin for future prediction calculations of moisture content in the construction elements of reconstructed walls with interior insulation.

1 INTRODUCTION

In order to reach the objectives of the Swiss Energy Strategy 2050 (1) it is necessary to improve the heat protection of the envelopes of the existing building stock in Switzerland including a high number of historical buildings.

To preserve the cultural heritage, walls of historical buildings have to be renovated with interior heat insulation systems.

Energetic refurbishments of walls in historical buildings with interior heat insulation show higher risks of damage caused by high moisture contents in the construction. Therefore, the robustness with regard to moisture protection of the construction is predicted by numerical programs, calculating the moisture content of the construction layers.

2 DESCRIPTION OF THE RESEARCH

Exterior walls of three historical buildings in Switzerland were refurbished with interior heat insulation. As insulation materials glass wool, stone wool and cellulose fiber were used. The three buildings are situated in Aarwangen, Brüttelen and in Bütschwil.

2.1 Property in Aarwangen

The first building is part of a farm which is located in Aarwangen. The building was erected in 1908.

The walls were renovated with an internal stone wool insulation, with a thickness of 12 cm.

Table 1 shows the parameters of the layers of the refurbished wall.

Table 1. Parameters of the layers of the refurbished masonry wall in the first object.

	Thickness in cm	Material
External plaster	ca. 1 cm	
External wall	ca. 30 cm	Brick
Thermal insulation	8 cm	Stone wool, thermal conductivity 0.035 W/(mK)
Vapor Barrier		Variable vapor barrier
Thermal insulation in installation cavity	4 cm	Stone wool, thermal conductivity 0.033 W/(mK)
Internal cladding	1.5 cm	Gypsum fiber board

2.2 Property in Brüttelen

The second building is situated in Brüttelen. The building was erected in 1900.

The walls were refurbished with an internal glass wool insulation, with a thickness of 17 cm.

Table 2 summarizes the parameters of the layers of the refurbished wall.

Table 2. Parameters of the layers of the refurbished wall in the second object.

	Thickness	Material
External plaster	ca. 2 cm	Cement plaster
External wall	ca. 50 cm	Stone
Thermal insulation	4 cm	Glass wool, thermal conductivity $\lambda = 0.035$ W/(mK)
Thermal insulation and vapour barrier	10 cm	Glass wool, thermal conductivity $\lambda = 0.035$ W/(mK) variable vapour barrier at the inner side $s_d = (0.3$ to $20)$ m
Thermal insulation in the installation cavity	3 cm	Glass wool, thermal conductivity $\lambda = 0.035$ W/(mK)
Internal cladding	1.5 cm	Gypsum fiber board

2.3 *Property in Bütschwil*

The third building is situated in Bütschwil. The building was erected in 1958.

The wall was refurbished with an internal cellulose fiber insulation, with a thickness of 14 cm.

Table 3 summarizes the parameters of the layers of the refurbished wall.

Table 3. Parameters of the layers of the refurbished masonry wall in the third object.

	Thickness	Material
External plaster	ca. 2.0 – 2.5 cm	
External wall	ca. 35 cm	Brick
Internal plaster	ca. 2.0 – 2.5 cm	
Thermal insulation	14 cm	Cellulose fiber, thermal conductivity $\lambda = 0.038$ W/(mK)
Internal cladding	1.5 cm	Gypsum fiber board

3 MEASUREMENT SETUP

The moisture contents and temperatures of the material layers in these reconstructed walls were measured on-site during a time period of two years.

In two buildings the measurement setup was installed at two positions in exterior walls at different directions to consider the influence of weather e. g. rain fall.

The measurement sensors are positioned in the external plaster, in the masonry wall and in the thermal insulation layer. As a measurement interval 10 minutes for the measurement period of two years were chosen. In addition, the temperature and the relative humidity of the air in the room behind the wall were measured. The exterior climate data was taken from the Swiss weather portal IDAweb (2).

4 LABORATORY MEASUREMENTS

To achieve a high conformity between simulation and measurement several laboratory measurements were done. Therefore, test specimens were taken from the walls. Among other things the laboratory measurements included the measurement of material parameters of the masonry and the exterior plaster, which were done by Fraunhofer Institute for Building Physics, Holzkirchen.

5 SIMULATIONS

The water content of different layers of the reconstructed walls were calculated by using a numerical program for transient coupled heat and moisture transport, named Wufi Pro (3).

In order to minimize the differences between the simulation and the measurement values a large number of calculation runs for the three objects were performed. To obtain minimal differences between simulation and measurement it was necessary to change the material parameters like the liquid transport coefficient of the masonry wall to consider the moisture transport via the mortar in the wall.

To reproduce the high value of water contents at the beginning of the measurement period, additional humidity sources had to be added in the simulations. By using these humidity sources penetrating water caused by leakages at windows could be considered.

6 RESULTS

The most critical position in the renovated walls is located between the masonry wall and the interior insulation. Therefor the differences between the simulation and measurement values of the temperatures and the relative humidity are presented for this position.

Because of a temporal fail of the sensor between masonry and interior thermal insulation in Bütschwil the results are presented for the position inside the insulation.

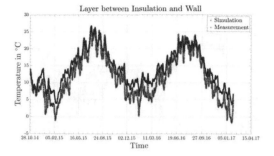

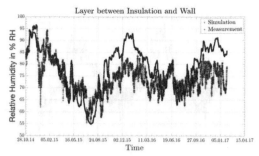

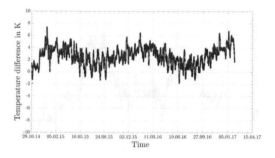

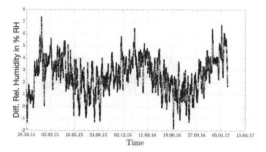

Figure 1. Comparison of the calculated and measured temperatures: the top part of the figure shows the simulated (small circles) and measured (big crosses) temperatures, the bottom part the difference between them, both as a function of time.

Figure 2. Comparison of the calculated and measured values of the relative humidity: the top part of the figure shows the simulated (small circles) and measured (big crosses) values of the relative humidity, the bottom part the difference between them, both as a function of time.

6.1 Property in Aarwangen

In this building measurements were carried out in two walls: the northwest and the southwest wall.

Figure 1 shows the comparison between the calculated and measured temperatures in the layer between the historic wall and the internal insulation in the northwest wall of the building.

The numerical values for the temperatures in the different layers of the renovated walls fit well to the measurement data.

Regarding the temperature of the layer between the insulation and the masonry wall the mean difference is (2.6 ± 2.8) K.

Figure 2 shows the comparison between the calculated and measured relative humidity in the layer between the historical wall and the internal insulation.

The deviation of the calculated values from the measured values of the relative humidity as the mean of the difference is (4 ± 13) % RH.

To achieve these small deviations, it was necessary to adjust the parameter set of the wall. In order to consider the influence of the mortar, the parameters of the brick wall are adapted as shown in Table 4.

6.2 Property in Brüttelen

In the second building measurements were carried out in two walls: the northwest and the southwest wall.

Table 4. Adapted material parameters of the northwest wall.

Vapor resistance μ		
Relative humidity	$0 - 0.50$	$0.5 - 0.93$
Measured value of the brick	26	23
Fitted value of the wall	17.8	17.6

Water absorption coefficient D_{ws} in m^2/s		
Relative humidity	0.80	1.0
Measured value of the brick	$2.5 \cdot 10^{-11}$	$8 \cdot 10^{-9}$
Fitted value of the wall	$2.8 \cdot 10^{-10}$	$2.5 \cdot 10^{-6}$

Water transport coefficient D_{ww} in m^2/s		
	0.80	1.0
Measured value of the brick	$3.9 \cdot 10^{-10}$	$2.5 \cdot 10^{-9}$
Fitted value of the wall	$6.5 \cdot 10^{-10}$	$1.1 \cdot 10^{-8}$

Water content u in kg/m^3						
Relative humidity	0	0.65	0.8	0.93	0.97	1
Measured value of the brick	0	27	39	69	109	235
Fitted value of the wall	0	7.7	8.2	9.7	13	335

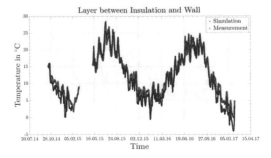

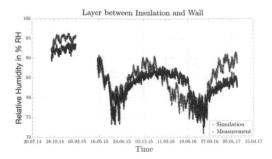

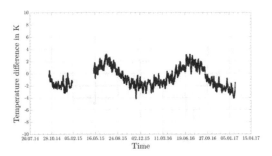

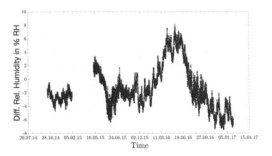

Figure 3. Comparison of the calculated and measured temperatures: the top part of the figure shows the simulated (small circles) and measured (big crosses) temperatures, the bottom part the difference between them, both as a function of time.

Figure 4. Comparison of the calculated and measured values of the relative humidity: the top part of the figure shows the simulated (small circles) and measured (big crosses) values of the relative humidity, the bottom part the difference between them, both as a function of time.

Figure 3 shows the comparison between the calculated and measured temperatures in the layer between the historic wall and the internal insulation in the northwest wall of the building.

Again, the numerical values for the temperature of the simulation and the measurement fit well: the mean deviation between simulation and measurement is (-0.8 ± 3.0) K.

Figure 4 shows the comparison between the calculated and measured values of the relative humidity in the layer between the historical wall and the internal insulation.

The deviation of the calculated values from the measured values of the relative humidity as the mean of the difference is (-1.3 ± 6.2) % RH.

To achieve these small deviations, it was necessary to adjust the parameter set of the wall. In order to consider the influence of the mortar, the water transport coefficient of the wall is adapted as shown in Table 5.

6.3 Property in Bütschwil

In the third building, measurements were carried out only in the north wall.

Figure 5 shows the comparison between the calculated and measured temperatures in the layer

Table 5. Adapted material parameters of the northwest wall.

Water transport coefficient D_{ww} in m^2/s		
Relative humidity	0.80	1.0
Measured value of the stone	$4.2 \cdot 10^{-9}$	$8.7 \cdot 10^{-8}$
Fitted value of the wall	$4.2 \cdot 10^{-10}$	$8.7 \cdot 10^{-9}$

inside the internal insulation in the north wall of the building.

Again the numerical values for the temperature of the simulation and the measurement fit well: the mean deviation between simulation and measurement is (0.4 ± 1.1) K.

The high deviation at the beginning and the end of the measurement period were due to the opening of the wall in order to change a sensor.

Figure 6 shows the comparison between the calculated and measured values of the relative humidity in the layer inside the internal insulation.

The deviation of the calculated values from the measured values of the relative humidity as the mean of the difference is (-0.2 ± 2.2) % RH.

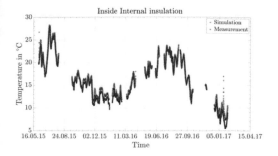

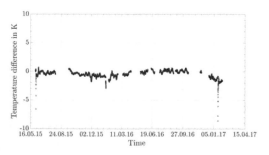

Figure 5. Comparison of the calculated and measured temperatures: the top part of the figure shows the simulated (small circles) and measured (big crosses) temperatures, the bottom part the difference between them, both as a function of time.

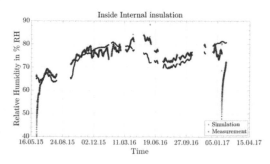

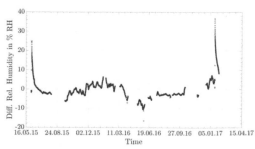

Figure 6. Comparison of the calculated and measured values of the relative humidity: the top part of the figure shows the simulated (small circles) and measured (big crosses) values of the relative humidity, the bottom part the difference between them, both as a function of time.

To achieve these small deviations, it was necessary to adjust slightly the parameter set of the materials.

Again the high deviations between simulations and measurement at the beginning and the end of the measurement period were caused by the opening of the wall construction.

7 CONCLUSIONS

By comparing the measurement values with the calculated ones, the reliability of the prediction calculation was detected for all three objects. The predicted values of the simulation program reproduce the measured values of moisture content and temperature in the construction layers within a small error margin.

To achieve a good agreement between measurement and simulation the parameter sets of the historical walls have to be adjusted to consider the humidity flow through the mortar. This was done in an individual manner for each wall. This is unsatisfactory, because general rules for this adjustment could not be identified. To identify these general rules more research has to be done.

Nevertheless, the differences between measurements and simulations can be interpreted as minimal deviations between simulation and reality. Thus, they may be used as a safety margin for future predictions of moisture content in the construction elements of reconstructed walls with interior insulation.

ACKNOWLEDGEMENT

We want to thank the Swiss Federal office for Energy and our industrial partners Flumroc AG, CH-8890 Flums, Isofloc AG, CH-9606 Bütschwil, Saint-Gobain ISOVER AG, CH-3550 Langnau and Pro clima Schweiz/MOLL bauökologische Produkte GmbH, D-68723 Schwetzingen for the funding of the project. Furthermore, we like to thank the owners of the buildings for the permission to install the measurement setups.

REFERENCES

Energiestrategie 2050 – Erstes Massnahmenpaket Swiss Federal office for Energy, 13.09.2012.
MeteoSchweiz, Datenportal IDAweb, Swiss Federal office for Meteorology und Climatology, 01.12.2014. Available http://www.meteoschweiz.admin.ch/home/service-und-publikationen/beratung-und-service/datenportal-fuer-lehre-und-forschung.html.
Wufi Pro, Fraunhofer Institut für Bauphysik, Holzkirchen.

Immediate measures to prevent further damage to the wall frescos of the "Ritterhaus Bubikon"

K. Ghazi Wakili & Th. Stahl
IABP, Institute for applied building physics, Winterthur, Switzerland

D. Tracht
Museum Ritterhaus Bubikon, Bubikon, Switzerland

A. Barthel
Canton of Zurich, Building Department, Conservation and cultural heritage, Dübendorf, Switzerland

ABSTRACT: The Ritterhaus Bubikon close to Zurich is considered as the best preserved commandery of the Order of St. John/Order of Malta in Europe. Its history dates back to the year 1192. Since 1959 the Ritterhaus has been under protection of the Swiss federal government. The east-oriented walls of the entrance hall to the unheated Ritterhaus chapel are covered with bacteria and algae caused by condensation during the spring time period. They have been recognized responsible for the partial destruction of frescos of those walls. By closing all openings of the entrance hall and the installation of a controlled ventilation system, condensation on the walls during spring time has been avoided. This was made possible by operating the ventilation to blow in the outdoor air only when its absolute humidity was lower than that of the indoor air during the critical period. The efficiency of the whole procedure has been proven by measured results of a simultaneous monitoring.

1 INTRODUCTION

The Ritterhaus Bubikon close to Zurich is considered as the best preserved commandery of the Order of St. John/Order of Malta in Europe (Figure 1). Its history dates back to the year 1192. The Society of the Ritterhaus Bubikon founded in 1936, saved the unique medieval building from downfall and established a museum about the house and the orders of the knights. Since 1959 the Ritterhaus has been under protection of the Swiss federal government.

In fact, the impressive building complex with its gray walls, ancient windows and picturesque fireplace seems to have been catapulted directly from the Middle Ages into our modern times. It has to be emphasized that without a steady maintenance and economically viable use, the complex would have remained but as a ruin. The approximately 800-year history of the Ritterhaus can be divided into four phases of use, which have left their specific traces:

- From the founding to the Reformation, the building complex served as commandery of the Order of St. John.
- As a result of the Reformation, the commandery lost its meaning. However, it remained the center of a manorial rule of the Order but was managed by the city of Zurich. Parts of the building complex were used as granary.

- Although the private, mainly peasant owners of the period 1789-1938 partly intensified the use of the building complex, they did not have the means for major modernization.
- With the acquisition of the whole building complex by the "Ritterhausgesellschaft" in 1938, the long-term preservation was secured. The restoration, carried out between 1938 and 1959, was done with the utmost care and respect for the historical fabric of the building. Nevertheless, it has to be kept in mind that every restoration is a child of its own time, Böhmer (2011).

2 RESEARCH OUTLINE

2.1 Initial state and improvement procedure

The object of investigation is the eastern wall of the entrance hall (dotted ellipse in Figure 2) where on its upper and lower parts pink and black discolorations are visible. It has been reported that mainly in springtime condensation water was dripping down the cold walls (no heating during winter time). The aim of the present investigation is to ameliorate the existing situation by monitoring the ventilation schedule dependent on the moisture contents of both the outdoor and the indoor air.

Figure 1. Overall view of the "Ritterhaus Bubikon". The dotted circle indicates the small investigated hall in front of the chapel.

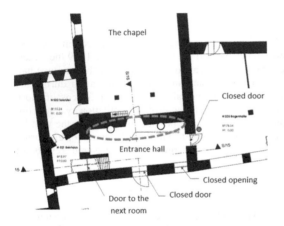

Figure 2. Plan view of the chapel and its entrance hall. The investigated wall with frescos is shown by a dotted ellipse. The openings which were closed for the present investigation are also indicated.

2.2 In-situ measuring equipment

For the present investigation, all openings of the hall have been closed (also indicated in Figure 2) and an electric fan installed at the bottom of the closed door facing the eastern wall (Figure 3). The fan was operated with a capacity of about 400 m³/h, which corresponds to about a double air exchange per hour.

Then a controlling unit was set up with a relay switch which opened and started the fan whenever the absolute moisture content of the outdoor air was lower than the absolute moisture content of the indoor air. Whenever this condition was not met the fan was shot down and closed so that the moist outdoor air was prevented from reaching into the entrance hall.

Six temperature sensors were placed on different parts of the wall and a pair of combined temperature and moisture sensors were installed in the room and respectively outside in front of the door to measure the conditions of the indoor and outdoor air. Data logging and regulation of the fan was done by an Ahlborn equipment ALMEMO®. Measured values

Figure 3. Installation of an electric fan at the bottom of the closed door facing the entrance to the chapel.

were stored every 6 minutes. In order to prevent a continuous triggering of the fan, a difference of 1 K was defined between the room side and the outdoor dew point temperature as triggering condition.

Two additional sensors were placed in 2 ducts of 80 cm depth just in front of the investigated wall (2 circular dots in Figure 2) to measure the absolute humidity at those sites. This would give a hint on the existence of rising damp. These ducts were dug in a previous project to investigate rising damp.

2.3 Evaluation of measured hygro-thermal data

Figure 4. shows the comparison of the absolute humidity of the outdoor and indoor air over the measuring period. Over and over again, there are overlaps of the two absolute moisture contents as well as longer periods during which the outdoor air is wetter than the indoor air (gray curve above the black curve in Figure 4). During these periods it is important not to let the outdoor air into the room.

The functionality of the electric fan in performing as expected is depicted in Figure 5. For matters of legibility, a period of 7 days (April 7th to April 14th) has been extracted from the whole measuring period of 3 months. The black and the gray lines represent the dew point of the indoor and outdoor air respectively. The additional 1 K shows the criterion for triggering the fan. The dark gray line at the bottom shows the "ON" (value = 1) and "OFF" (value = 0) positions of the fan. Figure 5 shows that whenever the dew point temperature of the outdoor air was higher than the dew point temperature of the indoor

126

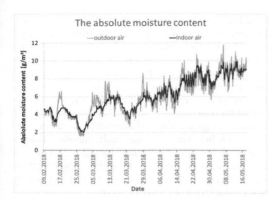

Figure 4. The measured absolute moisture content of the outdoor (gray) and the indoor (black) air during the investigation period.

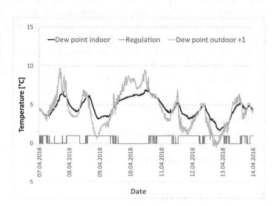

Figure 5. The dew point temperature of indoor air (black) and the outdoor air (gray) with the open and closed intervals for the fan (dark gray).

temperature the fan was closed and no outdoor air entered the room. By this it was possible to avoid moisture accumulation during the whole measuring period. It has to be reminded that the dew point temperature is an equivalent to the absolute moisture content of the respective air.

Comparing the measured absolute humidity of the indoor air and the air in the two ducts near the wall mentioned above, the possibility of rising damp can be excluded (Figure 6).

2.4 Evaluation of the microbiological analysis

Small surface probes with black and pink discolorations on them were taken from the wall and sent to the Fraunhofer Institute in Holzkirchen for analysis. The black discoloration was found to be due to a blue alga called Gleocapsa atrata KÜTZING. This needs liquid water (condensation) to survive. Preventing condensation on the wall surface means the destruction of the living conditions of this alga.

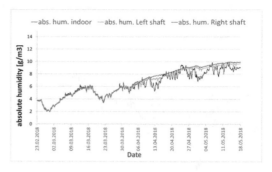

Figure 6. The absolute humidity of the indoor air and the air measured in the 80 cm deep ducts (circular dots Figure 2).

The pink discoloration is due to the presence of bacterial colonies of the type Micrococcus sp and its typical carotenoid pigmentation. It seemed that these colonies were already dead and unable to revive under laboratory condition. This might be due to a previous chemical treatment. These bacteria need a relative humidity level of 87-93 % at air temperatures between 20°C and 40°C. By keeping the air relative humidity beneath 87% a further growth can be ruled out. Figure 6 shows this for the whole measuring period (February 9th to May 16th 2018). The dark gray line representing the relative humidity of the indoor air remains below this threshold of bacterial growth. This can be even increased in future by increasing the air exchange rate of the fan so that a larger amount of dry air is blown into the room.

The relatively wide range of light gray in Figure 7 represents the average of the 6 measured surface temperatures of the wall under investigation. If now the temperature at surfaces are higher or lower than the temperature of the indoor air, the relative humidity at these surfaces will correspondingly be lower or

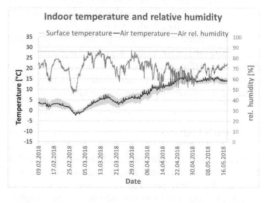

Figure 7. Evolution of indoor air temperature and relative humidity during the whole measuring period. The dashed gray line is the threshold in relative humidity for bacterial growth.

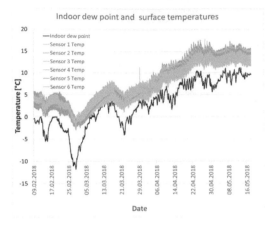

Figure 8. Surface temperatures compared to the dew point temperature of the indoor air.

higher than the relative humidity of the indoor air. However, if only a small temperature difference exists, the corresponding values of the relative humidity will not differ greatly from each other. That the latter is indeed the case is also shown in Figure 7.

For the present case March seems to be the most critical month. This depends of course on the location and the orientation of the building under investigation.

In a further consideration, it was investigated whether the measured surface temperatures fall below the dew point temperature of the indoor air. This would be an indication of condensation on the surface. As shown in Figure 8, this was not the case during the measurement period. The gray curves represent the 6 measured surface temperatures individually in comparison to the dew point temperature of the indoor air (black line). Nevertheless, some sensors came very close to the dew point temperature in mid-March. Thet fan power still has room to move upwards (max. 670 m³/h), so that this critical situation can be mitigated by a higher air eexchange rate.

The evaluation shows that in the mainly critical springtime the dew point was not undershot and there was difference of several Kelvin with respect to the dew point of the indoor air. This is a direct proof of the suitability of the action taken.

3 CONCLUSIONS

The room side air temperature of the entrance hall to the unheated Ritterhaus chapel was monitored and kept to the driest possible condition by means of a fan pushing outdoor air whenever its absolute humidity was lower than that of the room-side air. The measurements of the critical spring-time period show that it is possible to avoid condensation on the walls and re-duce the moisture content of the room side air. It was shown that by doing so it was possible to prevent the growth of algae and bacteria. This in turn stops the further deterioration of the remaining frescos.

The findings are also in-line with previous investigations on heritage buildings under similar climatic conditions Kiessl (1995) and Kilian (2012). The present can also be seen as a useful supplement to thermo-hygric quality in heated rooms containing protected artefacts Corgnati (2010).

There are some recommendations which would be helpful in the further investigations, especially when tackling the whole building complex:

All openings in the entrance hall must be closed to the outdoor by appropriate measures. In order to let enough natural light into the entrance hall, using glass doors is recommended. The installation of a suitable fan in the door is needed too.

A sensor-controlled ventilation control must be installed in the future if microbial contamination on the wall surfaces is to be avoided.

There is a necessity for control measurements of the indoor climate (proof of functionality and possibly necessary fine adjustments).

It is advisable to think about a control system with which the climate of the chapel and, if necessary, also of other adjacent rooms can be controlled.

ACKNOWLEDGMENTS

The technical and financial support of the "Society of the Ritterhaus" represented by Beat Meier is highly appreciated by all the authors.

REFERENCES

Böhmer R. 2011. Das Ritterhaus ein herausragendes Baudenkmal *Festschrift 75 Jahre Ritterhaus Bubikon* (in German), Switzerland.

Corgnati S.P. & Filippi M. 2010. Assessment of thermo-hygric quality in museum: Method and in-field application to the "Duccio Buoninsegna" exhibition at Santa Maria della Scala. *Journal of Cultural Heritage*, 11, 345–349.

Kiessl K. & Holz D. 1995. Klimaabhängige Belüftungssteuerung und einjährige Feuchtekontroll-Untersuchungen im Obergeschoss der Torhalle Lorsch. *IBP-Bericht FB-64/1995, Fraunhofer Institut für Bauphysik Holzkirchen* (in German), Germany.

Kilian R. & Kosmann, S. 2012. Torhalle Lorsch. Entwicklung eines Klimakonzeptes. *IBP-Bericht RK 008/2012/294, Fraunhofer Institut für Bauphysik Holzkirchen* (in German), Germany.

Preventive Conservation - From Climate and Damage Monitoring to a Systemic and Integrated Approach – Vandesande, Verstrynge & Van Balen (eds)
© 2020 Taylor & Francis Group, London, ISBN 978-0-367-43548-6

Energy retrofit of historic timber-frame buildings – hygrothermal monitoring of building fabric

C.J. Whitman, O. Prizeman & J. Gwilliam
Welsh School of Architecture, Cardiff University, Cardiff, UK

P. Walker & A. Shea
BRE Centre for Innovative Construction Materials, Department of Architecture & Civil Engineering, University of Bath, Bath, UK

ABSTRACT: In line with its aim to decarbonize the EU's building stock by 2050, the May 2018 amendment to the Energy Performance of Buildings Directive calls for "research into... the energy performance of historic buildings... while also safeguarding and preserving cultural heritage." To date in the UK, research in this field has focused on solid masonry construction. The research in this paper explores the previously under-researched retrofit of historic timber-framed buildings. In situ monitoring highlights that in some instances, the combination of incompatible materials, flawed detailing, poor workmanship and lack of controlled ventilation can facilitate biological attack. Digital hygrothermal simulations suggest that orientation, climatic conditions and infill material all significantly influence hygrothermal behavior, however, no prolonged periods of conditions favorable to biological decay were identified. Initial monitoring of test panels under laboratory conditions supports these results, however further long term monitoring is required and is currently underway, funded by Historic England.

1 INTRODUCTION

Energy retrofits have been identified as a key action to decarbonize the UK's building stock and improve hygrothermal comfort (DECC, 2014, OJEU, 2018). When undertaken with sufficient knowledge and consideration, the energy retrofit of historic buildings can be successfully achieved (Historic England, 2012). However, aesthetic, philosophical and technical issues must be fully understood in order to avoid unintended consequences (ibid.). As stated by the European Standard BS EN 16883 Conservation of Cultural Heritage - Guidelines for improving energy performance of historic buildings "[the] challenge is to reduce energy demand and greenhouse gas emissions without unacceptable effects on the heritage significance of the existing built environment" (British Standards Institution, 2017). To achieve this goal the Standard presents a systematic approach to facilitate the decision-making process (Figure 1). However, a key stage, mentioned in the Standard's text but not included in the original diagram, is the need for post-occupancy evaluation and feedback to close the loop. It is therefore essential for academic research to actively monitor and assess both current and future retrofit solutions for the historic built environment.

Research in the UK in this field has so far focused on the predominant solid masonry construction (Baker and Rhee-Duverne, 2015, Currie et al., 2013, Gandhi et al., 2012), with little research covering the 68,000 historic timber-framed buildings that form an integral part of the UK and specifically England's cultural identity (Whitman, 2017). This paper explores this previously under-researched area.

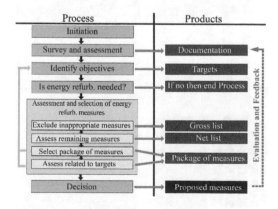

Figure 1. Flow chart showing procedure proposed by BS EN 16883. Source: based on (British Standards Institution, 2017) amended by author to include evaluation and feedback.

1.1 UK historic timber-framed construction

Archaeological evidence of timber construction can be found in the UK dating back to Neolithic times (Hillam et al., 1990). One of the oldest timber buildings still standing, the church of St. Andrews, Greensted-juxta-Ongar, Essex, dates from the late 11th to early 12th century AD. This building with its walls of solid half trunks is however not representative of the majority of timber buildings constructed in the UK from Mediaeval times until the late 18th century, which employed a timber frame with solid infill panels, the frame often exposed both internally and externally (Figure 2).

Infill panels were traditionally often of wattle-and-daub a framework of thin timber members (wattle-work) covered by an earthen render (daub). Other historic infills include lath and plaster and brick nogging (Harris, 2010). Where these historic infill materials survive, their conservation should be paramount. However, where they are beyond repair, have already been replaced with inappropriate materials or where their renewal is required due to conservation of the historic timber frame, there exists the opportunity to replace them with materials with improved thermal performance (Historic England, 2016).

A key concern with the energy retrofit of historic timber-framed buildings is the risk of elevated moisture content, increasing the potential for insect attack

Figure 2. 15th century exposed timber-frames, Church Street, Lavenham, Suffolk, UK. Source (Whitman, 2017).

and fungal decay (ibid). This paper presents research that begins to assess this risk with relation to the installation of replacement of infill panels.

2 DESCRIPTION OF THE RESEARCH

In order to explore the impact of replacement infill panels as an energy retrofit action for historic timber-framed buildings in the UK, a variety of methodological techniques were applied. These were, in situ monitoring, digital simulation and laboratory testing. There follows details of the methodologies employed and a summary of the results obtained.

2.1 In situ monitoring

In situ hygrothermal monitoring was undertaken at five historic timber-framed buildings in the UK. The case studies were selected to represent a variety of retrofit solutions, ownership models and uses (Table 1). Three of the case studies were located in Herefordshire (1-3) and two in East Anglia (4&5), both areas with a high concentration of the UK's surviving historic timber-frame buildings.

In summary, the retrofits of each case study are as follows; case study 1 had a mixture of replacement infill panel details, including some with traditional wattle-and-daub and others with a multi-foil insulation, in addition to underfloor heating with a ground source heat pump; case study 2 had increased roof insulation and secondary glazing, with no changes made to the external walls; case study 3 had replacement panel infills of woodfibre and woodwool following the detail published by Historic England (McCaig and Ridout, 2012 p.325); case study 4 had no retrofit but is undergoing 'conservative repair' in line with the ethos of the Society for the Protection of Ancient Buildings (SPAB); and case study 5 had replacement infill panels of rigid polyisocyanurate (PIR) boards finished with gypsum plasterboard, roof insulation and double glazing.

Monitoring included U-value measurements following BS ISO 9869-1:2014 (British Standards Institution, 2014), thermography following best practice guidance (Hart, 1991, Young, 2015), pressure testing according to BS EN ISO 9972:2015 (British Standards Institution, 2015), measurements of internal hygrothermal comfort using TinyTag Ultra 2 TGU-4500 sensors and

Table 1. Summary of case studies.

No.	Age	Use	Ownership	Retrofit
1	C15[th]	Holiday let	Private	Extensive
2	C16-19th	Dwelling	Charity	Partial
3	C17[th]	Commercial	Charity	Extensive
4	C14[th]	Dwelling	Private	None
5	C16[th]	Dwelling	Private	Extensive

Table 2. Comparison of U-values measured in situ according to BS ISO 9869-1:2014 and those calculated according to BS EN ISO 6946:2007. (Note: No in situ monitoring was possible at case study 2 due to personal circumstances of the resident).

Case Study	Panel build-up	Measured U-value (W/m²K)	Calculated U-value (W/m²K)	Difference Measured to calculated (W/m²K)
1.	Original lime plaster on oak lath	2.21	2.40	0.19
	New wattle-and-daub	2.88	2.99	0.11
	Multi-foil insulation	0.66	0.41	-0.25
3.	Woodfibre+ PIR internal lining	0.11	0.13	0.02
	Woodfibre+Mineral wool internal lining	0.11	0.17	0.06
4.	Pargeted with presumed wattle-and-daub infill	0.64	1.79	**1.15**
5.	Rigid PIR Insulation and Gypsum plasterboard	1.72	0.92	**-0.8**

simplified occupant questionnaires (Nichol et al., 2012). Timber moisture content was also monitored at two of the case studies (2 & 5) using electrical resistance measurements.

2.1.1 *Results*

The use of modern insulation materials should improve the thermal performance of the infill panels, however the measured U-values were often lower than those calculated (Table 2). This discrepancy can be attributed to the thermal bridging of the exposed timber-frame and poor detailing especially at the junction between frame and infill. At case study 4 the large discrepancy between measured and calculated U-values may be due either to incorrect assumptions regarding the wall build-up or the possible higher thermal performance of traditional building materials that has also been encountered by other researchers (Rye et al., 2012). However at case study 5, where the replacement infill detail achieved only 53% of the calculated U-value, this is principally due to poor design and installation of the replacement panel infill detail (Whitman et al., 2018b). Thermographic surveys (Figure 3) highlight the lack of hermeticity between infill and timber-frame, leading to a high air change rate and associated increased heat loss. This was confirmed through pressure testing, with an air permeability index of 19 m³/h.m².

Measurements of timber moisture content and temperature at the same case study showed that the use of non-vapour permeable materials, in conjunction with the poor detailing, had led to the creation of hygrothermal conditions that would facilitate biological attack, with one location providing conditions favourable to deathwatch beetle (*Xestobium rufovillosum*) for 99% of the monitoring period (Whitman et al., 2018b). Favourable conditions were also identified for house longhorn beetle (*Hylotrupes bajulus*), dry rot (*Serpula lacrymans*) and cellar rot (*Coniophora puteana*), although with lesser periods of duration (ibid.). The potential for fungal decay further increases the risk posed by insect attach as both the deathwatch and house longhorn beetles will only inhabit wood that has already been damaged by decay.

In case study 2, measurements pre and post-retrofit showed an increase in timber moisture content due to the retrofit actions increasing airtightness, without due regard to the provision of controlled ventilation. This highlights that whilst increased airtightness is required to improve these buildings' thermal performance, sufficient controlled ventilation must also be designed to manage internal relative humidity, especially in spaces with moisture sources such as bathrooms and kitchens.

The monitoring of internal hygrothermal conditions at all five case studies showed poor comfort conditions despite improvements to their external envelopes (Figure 4).

Influencing factors include poor airtightness, incoherent retrofit strategies and inadequate, inefficient heating. Despite this, the results of simplified occupant questionnaires and semi structured interviews at three case studies (1, 2 & 5) showed that occupants' thermal perceptions often contradicted the measured results, with their comments suggesting that their desire to live in these historic properties led to an acceptance of a lower thermal comfort threshold. This emotional response to heritage and its influence on occupant comfort and satisfaction presents an interesting area requiring further research.

Figure 3. Thermographic image of interior of case study 5, showing poor connection between infill and timber-frame.

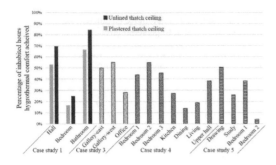

Figure 4. Percentage of inhabited hours measured where hygrothermal comfort was achieved.

2.2 Digital energy demand simulations

Digital simulations of the energy demand of the five aforementioned case studies were undertaken using the software DesignBuilder Version 4.2.0.54, a graphical interface for the dynamic simulation engine EnergyPlus DLL v8.1.0.009 (Design Builder, 2014). Climate files were created for each site using the software Meteonorm version 6.1 using the time period 1996-2005.

For each case study each of the individual retrofit actions were simulated separately in order to assess their specific impact on the building's heating demand. In addition, simulations of the combined effect of multiple retrofit actions, both those applied in reality and hypothetical scenarios, thereby allowing the assessment of the current and future potential performance of these buildings.

2.2.1 Results

For case study 1 the results showed that the mixture of replacement panel infills as built has little effect on the energy demand with only a 3% reduction as compared to a hypothetical scenario of all the infill panels being the surviving lime plaster on oak lath. In reality many of the original panels had already been lost or required replacement due to repair work to the surrounding timber frame. However, if this had not been the case, historic fabric could have been lost for little gain in energy efficiency. The results did however show that the simple act of lime plastering the previously unlined exposed thatch ceiling, reduced the air change rate by approximately 50%, resulting in a significant impact on the energy demand with a reduction of 36%.

At case study 2, the introduction of secondary glazing resulted in a reduction of 10% in the heating energy demand by simultaneously improving the thermal performance of the fenestration and the airtightness. This when combined with the roof insulation resulted in an overall reduction of 34%. The simulation of a hypothetical scenario replacing all the current infill panels (predominantly 20th century concrete block, with some surviving wattle-and-

daub) showed only a 9% improvement. Given this minimal improvement, it is questionable if this should be attempted in the future due to the major disruption to the building's occupant and the potential loss and damage to the historic fabric.

The simulations for case study 3 again indicated a limited impact of the replacement of the infill panels with woodfibre and woodwool, in conjunction with additional partial internal lining, with a reduction in heating energy demand of 12%. If instead 20mm of woodfibre insulation had been installed above the 17th century plastered ceiling then a reduction of 17% could have been achieved. It is understandable the conservation architect's reluctance to undertake work directly related to this element which is perhaps the building's most significant heritage asset. However, thermography undertaken during the in-situ monitoring highlighted large differences between the now insulated walls and the uninsulated ceiling, raising concern over the potential concentration of condensation on the latter. This highlights the complex and difficult decisions faced by those engaged in the energy retrofit of historic buildings and sustainable building conservation.

For case study 4 a heating demand of 113 kWh/m² was simulated. This had increased from 97kWh/m² in 2012 when the house had previously been pressure tested (Hubbard, 2012). The deterioration in airtightness had resulted from the removal of impermeable 20th century internal finishes as part of the conservative repair the building is currently undergoing. It is hoped that when this work is completed and new more appropriate finishes have been installed that the airtightness will improve and hence result in a reduction in the energy demand. Assuming no improvement in airtightness, a 25mm layer of woodfibre insulation applied as internal lining to the walls would see a potential reduction of 12%. However, there is concern over the impact of this insulation on the elaborate 17th century exterior pargetting, probably the buildings most significant feature. Modelling using THERM® version 7.5 showed that the temperature of the external surface of the pargetting would drop by 0.5°C with the introduction of the insulation, thereby increasing the risk of frost damage. Potentially a more appropriate solution would be to insulate the roof and exposed floors which combined could result in a reduction of 14%.

Finally, for case study 5, due to the previously mentioned poor thermal performance of the badly designed and installed replacement infill panels, it is possible that the heating energy demand increased by 1% as a result of the retrofit. It may however be reasonable to suppose that the original lath and plaster infill panels provided a more hermetic seal between panel and timber-frame and so possibly the increase in heating demand may be even higher. The current owner and their architect now intend to replace the rigid PIR insulation and gypsum plasterboard with sheep's wool insulation held between lath and plaster. It is hoped that this will result in a heating demand of 63 kWh/m^2 a reduction of 32% from the current situation.

Both the in situ measurements and the energy demand simulations demonstrate that if well designed and installed, the thermal upgrading of the walls of timber-framed historic buildings can be beneficial but only when considered as part of a whole house approach, with improvements to airtightness seeing the greatest impacts. Conversely, when poorly designed and installed, replacement infill panels have the potential to not only reduce the energy efficiency of the building but also put its historic fabric at risk.

2.3 Digital hygrothermal simulations

In order to study the risk of increased moisture content arising from replacement panel infills, digital hygrothermal simulations with WUFI® Pro 5.3 were undertaken. Thirteen replacement panel infill details proposed by current guidance (Historic England, 2016, Reid, 1989, McCaig and Ridout, 2012) were simulated in six geographical locations (Suffolk, Cambridgeshire, Kent, Devon, Herefordshire and Cumbria (Figure 5)), representing the principal climates where timber-framed buildings are to be found in England (Whitman, 2017).

2.3.1 Results
The results suggest that orientation, climatic conditions and infill material all significantly influence the moisture content, however, no prolonged periods of hygrothermal conditions favourable to biological decay agents were identified (Whitman et al., 2015). Those instances of favourable conditions that did occur were sporadic and limited to fewer hours than those required for the gestation of both insects and fungi.

The orientation with the highest risk was south. Although it had been assumed that the prevailing patterns of wind driven rain would prejudice a south west orientation, the increased exposure to direct solar gain of panels facing due south would appear to be more influential in creating the warm damp conditions favourable to biological attack.

Surprisingly the location with the highest risk was not that with the highest rainfall, Cumbria, but rather Suffolk, where higher rainfall occurs during summer months, coinciding with warmer temperatures. Although there currently exists some debate as to whether climate change will lead to increased summer rainfall, overall the predictions point to lower precipitation levels in the summer (Environment Agency, 2014), however, there are suggestions that there may be an increase in hourly rainfall intensities due to convection induced precipitation (thunderstorms) (Kendon et al., 2014 p.570). If so then the climate seen in Suffolk may become more common across the country, thereby increasing the risk of biological attack. This is an interesting area for further research.

The replacement panel infill detail with the highest risk of biological attack would appear to be hemp-lime due to the high initial moisture content of this construction. The drying time for built in moisture can be significantly reduced if work takes place at the beginning of the summer. This highlights the need for informed programming of such construction work.

It must however be acknowledged that these simulations represent idealised constructions with homogenous layers, rather than the heterogeneous reality, and that material data are limited for historic materials.

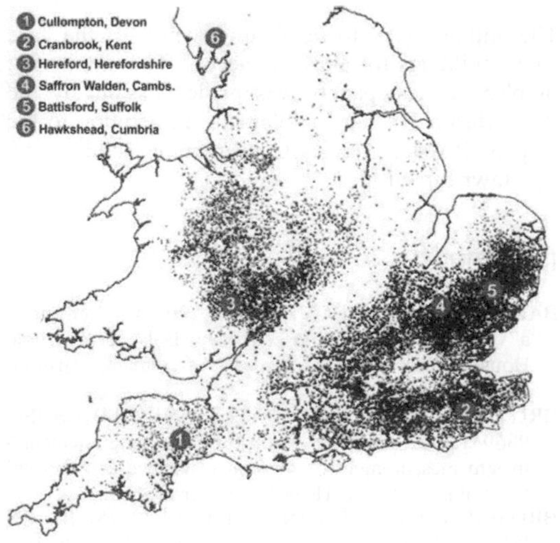

2.4 Physical test panels

Given the limitations of the digital simulations, the interstitial hygrothermal monitoring of physical test panels was also undertaken (Figure 6).

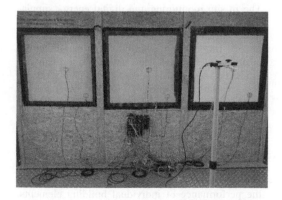

Figure 5. Geographical locations used for WUFI® Pro5.3 simulations of replacement infill panel details.

Figure 6. Panels in climate chamber. View from "internal" chamber.

Three test infill panels 1020mm x 1020mm x 100mm (L x W x D) were constructed within reclaimed oak frames. The dimensions were defined following a review of a representative sample of 100 historic timber-framed buildings in the UK. Dynamic Vapour Sorption (DVS) undertaken on oak samples felled in the 17th, 19th and 21st centuries showed that age did affect vapour sorption with the older samples absorbing less moisture (Demonstration & Contract Testing Services, 2015). Reclaimed oak was therefore used for the panel frames.

Following a review of details proposed by current guidance, the chosen infill materials were wattle-and-daub, expanded cork, and a detail using wood fibre and wood wool as suggested by Historic England (McCaig and Ridout, 2012). The panels were mounted as part of a dividing wall between two climate-controlled chambers at the University of Bath's Building Research Park.

Temperature and moisture content were monitored in the centre of the panel and at the interface between infill and oak frame at depths of 10mm, 50mm and 90mm. Results were compared with digital simulations using WUFI® Pro 5.3 and WUFI 2D using measured climatic data.

2.4.1 *Results*
Under extreme steady state conditions, sustained for three weeks, interstitial condensation was measured in the woodfibre/woodwool panel (Whitman et al., 2018a). However this did not reoccur during a following fortnight of cyclical dynamic conditions (ibid.). Whilst the digital simulations did successfully predict this condensation, discrepancies were identified both between measured and simulated data and between simulation methods. Further long term monitoring is now being funded by Historic England.

3   CONCLUSIONS

- The in situ monitoring and digital energy simulations have shown that modern insulation has the potential to improve the thermal performance of historic timber-framed buildings, although thermal bridging by the frame and poor detailing can significantly reduce their effectiveness. Those retrofit actions with the greatest impact on reducing energy demand were related to improving airtightness. However, these must always be undertaken whilst providing adequate controlled ventilation to avoid the increase in internal relative humidity and moisture content of internal finishes as seen in case study 2.
- Whilst retrofit solutions were shown to improve the performance of individual building elements, there was often a negligible increase in overall energy efficiency and hygrothermal comfort

conditions. However, the measured conditions did not always correlate with occupants' thermal perceptions. This may suggest a possible positive influence of the occupants' emotional connection to the buildings. This presents an interesting area for future research.
- The in situ monitoring at case study 5 highlighted that the use of non-vapour permeable materials and poor detailing can increase the risk of biological attack. This was confirmed by the digital hygrothermal simulation which concluded that panel orientation, climatic conditions and infill material all significantly influence the moisture content, with warm damp climates being most at risk. Given the possible increase in convection induced summer precipitation and warmer summer temperatures, this is an area requiring further research.
- The monitoring of physical test panels showed that under extreme conditions interstitial condensation has been observed to occur, however discrepancies exist between simulated and measured data. Further research over a longer period is therefore required. A test cell that will allow the monitoring over a minimum of two years, of four panel infill details, is currently under construction funded by Historic England.
- Together, the research presented in this paper demonstrates the complexities of the energy retrofit of historic timber-framed buildings and the need for monitoring and simulation to inform the decisions of those undertaking this work. By doing so it is possible to both learn from past mistakes and achieve the best outcomes with the minimal loss of historic fabric.

ACKNOWLEDGEMENTS

The authors wish to thank the owners of the case study buildings for allowing access. The monitoring of physical test panels was made possible by the APT Martin Weaver Scholarship, in addition to the help of Royston Davies Conservation Builders and Ty Mawr Lime Ltd.

REFERENCES

BAKER, P. & RHEE-DUVERNE, S. (2015) A Retrofit of a Victorian Terrace House in New Bolsover a Whole House Thermal Performance Assessment. Historic England.

BRITISH STANDARDS INSTITUTION (2014) BS ISO 9869–1:2014 Thermal insulation- Building elements- in situ measurement of thermal resistance and thermal transmittance Part 1: Heat flow meter method.

BRITISH STANDARDS INSTITUTION (2015) BS EN ISO 9972:2015 Thermal performance of buildings - determination of air permeability of buildings - fan pressurization method. British Standards Institution.

BRITISH STANDARDS INSTITUTION (2017) BS EN 16883 Conservation of Cultural Heritage - Guidelines for improving energy performance of historic buildings. BSI.

CURRIE, J., WILLIAMSON, J. B. & STINSON, J. (2013) Technical Paper 19: Monitoring thermal upgrades to ten traditional properties. Glasgow, Historic Environment Scotland.

DECC (2014) UK National Energy Efficiency Action Plan London.

DEMONSTRATION & CONTRACT TESTING SERVICES (2015) Customer Report- Samples WSA17, WSA19 & WSA21. Surface Measurement Systems Ltd.

DESIGN BUILDER (2014) Design Builder Help 4.2.

GANDHI, K., JIANG, S. & TWEED, C. (2012) Field Testing of Existing Stone Wall in North Wales Climate. *SusRef: Sustainable Refurbishment of Building Facades and External Walls*. Cardiff University.

HARRIS, R. (2010) *Discovering Timber-Framed Buildings*, Oxford, UK, Shire Publications.

HART, J. M. (1991) Practical guide to infra-red thermography for building surveys. Bracknell, UK, Building Research Establishment.

HILLAM, J., GROVES, C., BROWN, D., BAILLIE, M., COLES, J. & COLES, B. (1990) Dendrochronology of the English Neolithic. *Antiquity*, 64, 210–220.

HISTORIC ENGLAND (2012) Energy Efficiency and Historic Buildings: Application of Part L of the Building Regulations to historic and traditionally constructed buildings (Revised 2012). IN ENGLISH HERITAGE. (Ed. *Online*. UK, English Heritage.

HISTORIC ENGLAND (2016) Energy Efficiency and Historic Buildings: Insulating Timber-Framed Walls. IN HISTORIC ENGLAND (Ed.

HUBBARD, D. (2012) Air permeability testing and thermographic survey: 27 Church Street, Saffron Walden., ArchiMetrics.

MCCAIG, I. & RIDOUT, B. (2012) *English Heritage practical building conservation- Timber*, London; Farnham, Surrey; Burlington, VT, English Heritage; Ashgate.

NICHOL, F., HUMPHREYS, M. & ROAF, S. (2012) *Adaptive thermal comfort: Principles and practice*, Abingdon, Oxfordshire, UK, Routledge.

OJEU (2018) DIRECTIVE (EU) 2018/844 OF THE EUROPEAN PARLIAMENT AND OF THE COUNCIL of 30 May 2018 amending Directive 2010/31/EU on the energy performance of buildings and Directive 2012/27/EU on energy efficiency. Official Journal of the European Union.

REID, K. (1989) Panel Infillings to timber-framed buildings. IN SOCIETY FOR THE PROTECTION OF ANCIENT BUILDINGS. (Ed. London, UK.

RYE, C., SCOTT, C. & HUBBARD, D. (2012) THE SPAB RESEARCH REPORT 1. U-Value Report. Revision 2 ed., Society for the Protection of Ancient Buildings.

WHITMAN, C. J. (2017) The distribution of historic timber-framed buildings in the UK and the impacts of their low energy retrofit. Cardiff University.

WHITMAN, C. J., PRIZEMAN, O., GWILLIAM, J., SHEA, A. & WALKER, P. (2018a) Physical Monitoring of Replacement Infill Panels for Historic Timber-Framed Buildings in the UK. *Passive and Low energy Architecture (PLEA) 2018*. Hong Kong.

WHITMAN, C. J., PRIZEMAN, O., GWILLIAM, J. & WALKER, P. (2018b) The impact of modernization of a 16th century timberframed farmhouse, Suffolk, UK. *Energy Efficiency in Historic Buildings (EEHB) 2018*. Visby, Sweden.

WHITMAN, C. J., PRIZEMAN, O. & WALKER, P. (2015) Interstitial Hygrothermal Conditions of Low Carbon Retrofitting Details for Historic Timber-framed Buildings in the UK. *Passive and Low Energy Architecture (PLEA)*. Bologna.

YOUNG, M. (2015) Thermal Imaging in the Historic Environment. *Short Guide*. Historic Environment Scotland.

Preventive Conservation - From Climate and Damage Monitoring to a Systemic and Integrated Approach – Vandesande, Verstrynge & Van Balen (eds)
© 2020 Taylor & Francis Group, London, ISBN 978-0-367-43548-6

3D Laser scanning for FEM-based deformation analysis of a reconstructed masonry vault

A. Drougkas
Faculty of Civil Engineering and Geosciences, TU Delft, Delft, The Netherlands

E. Verstrynge
Building Materials and Building Technology Division, KU Leuven, Leuven, Belgium

M. Bassier & M. Vergauwen
Geomatics Research Group, KU Leuven, Ghent, Belgium

ABSTRACT: Reconstruction of historic building elements is often necessary in adaptive re-use projects. Optimally this is performed with as much original material as can be salvaged. However, the use of hydraulic lime mortars with no cement content in reconstructed masonry can lead to long curing time and excessive deformation under mechanical loads. Therefore, local masonry reconstruction in adaptive re-use projects using historic materials that need to adhere to pressing construction schedules should always be closely monitored.

The objective of the paper is to demonstrate the need for accurate geometric survey of vault structures in order to achieve accurate deformation results using numerical analysis. Focusing on a complex reconstruction project involving a masonry vault at the Royal Academy of Fine Arts in Ghent, practical aspects of damage monitoring, geometric survey and computational analysis of historic structures are jointly presented and addressed.

The vault was dismantled and reassembled using the original bricks and a newly made hydraulic lime mortar, the latter of which was mechanically characterized. Existing cracks in the masonry walls supporting the vault were monitored for the detection of new damage. Detailed geometric surveys were carried out using terrestrial laser scanning at two points in time after the reconstruction of the vault: a) before the removal of the formwork and b) after the removal. These scans are able to not only register the geometry of the vault in great detail, but also to measure the deflection of the structure under its self-weight non-intrusively and with good accuracy.

Structural analysis of the vault has been carried out employing two approaches: a) by using simple geometric models of the vault and b) by using the detailed laser scanning data. Major differences between the two approaches are obtained in terms calculated deflection, highlighting the importance of detailed geometric survey for the analysis of historic structures. Detailed geometric survey data is shown to be critical in achieving accurate analysis results in structures whose deformation behaviour is mainly governed by their geometry.

1 INTRODUCTION

Changes in use of historic buildings, particularly those having already suffered structural damage, may require partial reconstruction and/or strengthening of structural elements for the accommodation of increased loads or their redirection to sturdier parts of the structure. The use of original, where available, and compatible, where practical, materials is advisable.

Even when original or compatible materials are used, partial reconstruction of structural elements entails their unloading and subsequent reloading. The reconstruction and unloading/reloading process may induce changes in the load distribution within the building. It can therefore be required to monitor the behavior of the building during the reconstruction process in terms of deformation and damage propagation (Verstrynge, Schueremans, & Smars 2012).

Structural performance is not evaluated only in terms of stress or safety factors at the ultimate limit state. Buildings need to satisfy criteria related to limits on deformations under specific load conditions. Therefore, both the ability to reliably monitor movement in the structure, as well as to accurately predict its deformation under mechanical load, are critical for the assessment of historic structures.

Masonry structural elements such as arches and vaults derive their strength and stiffness from their shape (Milani, Milani, & Tralli 2008), both of which may be compromised by excessive deflection or movement of their supports. This is in contrast with elements such as walls and pillars, whose function is typically more straightforward in the case of vertical loads. Therefore, in structural analysis efforts, arches and vaults require special care for the accurate representation of their geometry to a degree that may not necessarily be required for walls and pillars.

Ideally, the results of material testing, geometric survey and monitoring feed into structural analysis techniques. These techniques, when properly set up, are mutually supportive and can greatly assist the assessment process for complex historic structures during and after construction (ISCARSAH 2003).

In this paper, a case study involving the reconstruction of a masonry vault is presented. For this project material testing, digital geometric survey, crack monitoring and finite element analysis were combined for the assessment of the behavior of the vault during its loading under its self-weight. Different geometric models of the vault are analyzed, illustrating the importance of digital survey for the dual purpose of deformation monitoring and geometric data acquisition to be used in numerical analysis.

2 THE CASE STUDY

2.1 *The vault structure*

The case study involves the building of the Royal Academy of Fine Arts in Ghent, Belgium. As is common in large historic structures, the building complex was constructed in phases over an extended period. The west wing was completed in 1612, the north wing in 1827 and the east wing in 1871. The overall layout of the building complex is shown in Figure 1.

The studied vault is a masonry cross-vault located on the top floor of the north wing. The original vault was constructed using solid clay bricks and lime mortar. It horizontally spans roughly 9.66 m × 8.00 m and has a rise of 1.23 m (Figure 1), giving the vault a rather shallow shape. Its thickness in the mid-section of the span, measured manually through an opening in the vault, is 0.20 m. This thickness corresponds to the length of the brick units used in the construction of the vault. The vault features a brick masonry fill near its four main vertical supports, constructed using a mostly regular bond rather than rubble fill material. Due to the

shallowness of the vault, this fill is not very massive, but still contributes to the weight and stiffness of the structure.

The vault after reconstruction may be seen in Figure 2. Of particular note is its low curvature and, therefore, low rise, resulting in a rather shallow construction. Also visible in the image are the masonry fills at the corner supports of the vault.

The plan of the vault and its location relevant to the other vaults of the building is shown in Figure 3. To the northern and southern side of the vault as seen in the plan, the vault is not counterbalanced by equally large vaults. Towards the right is located the outer wall of the building. Existing and new metallic ties were installed along the length of the vault's edge from left to right in the plan view. During reconstruction, an additional externally mounted steel beam was temporarily placed in this outer boundary, further limiting possible lateral movement towards the right. A similarly sized vault is located towards the left side of the studied vault, providing a counterbalancing action.

Due to cracking in the underlying masonry walls as well as in the anchoring location of the existing metallic ties, the vault was deformed and cracked. Hence, a decision was made for it to be dismantled and reconstructed using the original bricks and a new lime-based mortar. The reconstruction took place atop a rigid formwork, densely underpinned by scaffolds. A new reinforced concrete floor slab will be constructed over the vault and supported on the masonry walls, leaving the reconstructed vault to bear only its self-weight and a light infill.

2.2 *Material testing*

Samples of the new lime mortar were cast at the time of construction. Rigid plastic cylindrical containers were used for this purpose. The mortar in the containers was subsequently cut in 40 mm cubic samples and tested in compression at the ages of 14 and 21 days. Three samples were thus extracted for testing at each designated age. The tests were

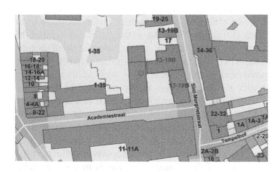

Figure 1. Layout of the royal academy of fine arts building in Ghent. North wing in blue.

Figure 2. View of the upper surface of the vault.

Figure 3. Plan of vault structure in red with position of DEMEC gauges and crack meters illustrated in blue (solid lines indicate placement above the vault, dashed lines indicate placement below the vault.

Table 1. Results of mortar compressive tests: mean and characteristic compressive strength.

Age [days]	$f_m[N/mm^2]$	CoV [−]	$f_k[N/mm^2]$
14	3.38	0.13	2.64
21	4.20	0.04	3.94

executed in displacement control, at a rate of 2 mm/min measured at the stroke of the press. With the exception of the production of the mortar samples through cutting, the tests were overall executed according to the requirements of the relevant European standard (CEN 2005). The results are presented in Table 1.

The average compressive strength f_m of the mortar at only two weeks after casting is already substantial. A noticeable increase of this compressive strength is registered for the samples aged 21 days. This age coincides with the time between finalization of the construction of the vault and removal of its formwork. It is therefore similar to the age of the on-site mortar when the vault's thrust line is activated. More pronounced is the decrease in the coefficient of variation (CoV) of the compressive strength of the mortar. Calculating the characteristic compressive strength of the mortar f_k as the 5% percentile based on a normal distribution of the test results, a significant increase is obtained in this parameter at 21 days. The results of the compressive tests, in terms of press stroke and stress, are illustrated in Figure 4.

2.3 Deflection and damage monitoring

A certain amount of deflection of the vault was expected due to the use of lime mortar, which can present, but also more efficiently accommodate, deformation in masonry. The need for early removal of the formwork due to the construction schedule, meaning that the structure would be loaded after only three weeks of setting time for the mortar, could potentially increase possible deflection of the vault under its self weight.

The deflection of the vault during the gradual lowering of the formwork was measured at specific points by the contractor using a total station. The formwork was carefully lowered at 10 mm increments separated by 30 minute intervals. The maximum deflection measured at the centre of the vault was 30.9 mm.

The ratio of the minimum span of the vault over this deflection is roughly equal to $30.09/8000 = 1/267$. Indicatively, this ratio satisfies by a small margin the limit of span/deflection ratio of $1/250$ for immediate elastic deflection under quasi-permanent loads specified by the Eurocode 2 for reinforced concrete floor slabs (CEN 2004). It should, however, be noted that the vault will not function as a load bearing floor and any additional deflection would not result in serviceability issues but only to the potential disturbance of the ceiling finish.

A number of demountable mechanical strain (DEMEC) gauge measurement points and crack meters were installed at visible cracks on the masonry walls supporting the vault, as well as on one of the old metallic ties over the vault. The cracks were located in lintels above doors in the floor below the vault. While the initiation of the cracks cannot be conclusively linked with the weight of the overlying vault, the relation between their further opening and the reactivation of the vault after the removal of the formwork was deemed critical. The layout of the gauge placement on the floor plan is illustrated in Figure 3.

Measurements were registered before and after the removal of the formwork and are given in Table 2. The monitoring data did not indicate any important opening of cracks or additional displacements. There were, however, small differences between the two locations of the DEMEC gauges on the underlying masonry walls. A1 and A2 gauges registered larger opening (between 0.192 and 0.439 mm) compared to the A3 and A4 gauges (which registered a nearly negligible closing of the cracks at the measurement location). These measurements indicate the better horizontal support in the direction of the A3 and A4 gauges, provided by the existing and newly placed metallic ties, the externally mounted steel beam as well as by the neighbouring large vault (Figure 3).

The B1 and B2 measurements indicate that the old metallic tie has been rendered inactive by the placement of a new passive anchor at the same loca-

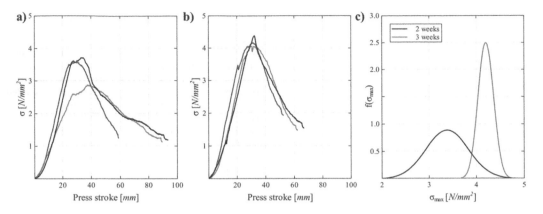

Figure 4. Results of mortar compressive tests: a) stress-displacement curves at 14 days, b) stress-displacement curves at 21 days, c) probability density function $f(\sigma_{max})$ for normal distribution of experimentally derived peak stress σ_{max} results.

Table 2. DEMEC gauge results before and after the removal of the formwork.

Gauge	Before removal [mm]	After removal [mm]	Difference [mm]
A1	-0.106	0.333	0.439
A2	0.478	0.670	0.192
A3	3.638	3.630	-0.008
A4	-1.984	-1.876	0.108
B1	-0.913	-0.903	0.010
B2	2.127	2.140	0.013
Temperature	20°C	28°C	

tion. Finally, the external temperatures registered during acquisition are also shown in Table 2. Of note is the significant increase in the external temperature (8°C) between measurements, which could potentially have masked some of the crack movements.

Alongside the DEMEC gauges in the two crack locations, tell-tale crack meters were placed on the masonry walls. The measurement accuracy of these crack meters is limited to 0.5 mm, below the magnitude of crack opening measured by the DEMEC gauges. Therefore, the crack meters did not register any measurable crack opening.

3 LASER SCANNING

3.1 Acquisition methodology

The deflection of the vault was determined through the use of point cloud data. A terrestrial laser scanner captured the geometry of the structure before and after the removal of the formwork. In this project, a Leica P30 camera was utilized, yielding a point cloud accurate up to 2 mm with a spatial resolution of 2 mm. The deflection was established by evaluating the cloud-to-cloud distance between both point clouds in the vertical direction. To avoid outliers, deformations smaller than 2 mm were not considered. This evaluation was executed using the software Cloud Compare (Girardeau-Montaut 2015).

For the first acquisition, 14 scans were taken on the floor above the vault and 11 scans underneath, amounting to roughly 600 million points. For the second acquisition, 9 scans were taken from above the vault, amounting to roughly 350 million points. The target area at the vault contained roughly 3 million points. Each scan required roughly 3-5 minutes for acquisition, meaning that each acquisition session lasted approximately half a working, engaging a crew of two surveyors.

 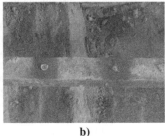 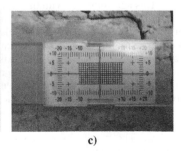

a)   b)   c)

Figure 5. a) Crack above door at the location of A3 and A4 DEMEC gauges. b) B2 DEMEC gauge points placed on existing tie. c) Crack meter after reloading of the vault.

Figure 6. Point cloud data distance in [m] between consecutive scan epochs in the scanned floor.

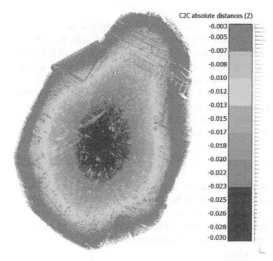

C2C absolute distances (Z)

Figure 7. Cloud-to-cloud distance in vertical direction between two scan epochs in the target area in [m] (plan view).

3.2 Scanning results

The results of the cloud comparison, indicating the movement of the vault compared to the remaining elements of the floor are illustrated in Figure 6.

The results of the scanning are shown in detail in Figure 7, where the area of the vault is isolated and the deformation profile is clearly seen. A maximum vertical displacement of 30 ± 2 mm is observed at the center of the vault. Horizontal displacements were negligible. The profile of the deflection is accurately determined and clearly defined in contrast to the areas that may be considered as stable. The magnitude of the scan-measured deflection is in good agreement with the manually measured deflection at the center of the vault.

In addition to determining the deflection of the vault, the point cloud acquired before the removal of the formwork was used to create a high-accuracy mesh model of the structure. The top side of the structure was reconstructed using Poisson meshing

(Kazhdan, Bolitho, & Hoppe 2006). Due to occlusion by the formwork and its scaffolds, direct acquisition of the lower side of the vault was not possible. The lower side was therefore generated by a fixed offset, equal to the vault thickness, from the topside. Any remaining openings in the mesh were manually sealed, resulting in a watertight solid model that served as the basis for part of the numerical modelling. The scan-derived finite element mesh was constructed using the software Gmsh Geuzaine & Remacle 2009).

Overall, laser scanning, was able to produce high detail geometric data with minimal acquisition time and no disturbance to the construction works and despite the presence of occluding scaffolds and the formwork.

4 NUMERICAL MODELLING

4.1 Model geometry

Although the capacity of arches and vaults is regularly calculated with good accuracy using graphic statics or limit analysis approaches (Como 2012), the current problem is not one of capacity but one of deflection. The vault will not bear live loads but will be loaded only under its self-weight. Graphic statics and limit analysis cannot provide deformation output in this context. A finite element approach is adopted instead for the structural analysis of the vault. This approach is capable of calculating the response of the structure in elastic bending.

The finite element mesh is determined in three ways. These are, in increasing complexity: a) an idealized shell element model, b) an idealized solid element model, c) a scan-derived model with solid elements. The two idealized meshes were constructed from primitives defining a cross vault with the spans and rise of the vault as measured from construction drawings and on-site manual measurements. This simple parametric approach for the geometry only considers the spans in the two directions, the vault thickness and the rise, and does not consider variations in the masonry thickness, the effect of imperfections and the presence of the fill. This approach disregards the influence in terms of added weight and stiffness of the masonry fill atop the vault near its main vertical supports. These models are shown in Figure 8.

All geometric features are directly considered in the scan-derived model, which incorporates fine details of the geometry of the vault, as shown in Figure 9.

Overall, the shell model has 9861 nodes and 4834 6-node shell elements, the solid model has 20501 nodes and 4530 10-node tetrahedral and 13-node wedge elements and the scan-derived model has 27384 nodes and 102322 4-node tetrahedral elements. While quadratic elements were used for the idealized meshes, linear elements were used for the scan-derived model. This was done for the sake of maintaining the

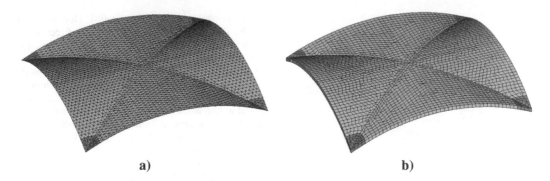

a)   b)

Figure 8. Simple finite element meshes used for analysis: a) shell model and b) solid model.

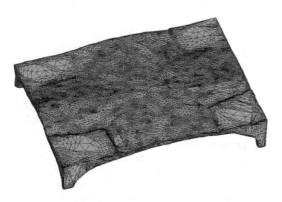

Figure 9. Scan-derived mesh for finite element analysis.

geometric fidelity of the scan through the use of small elements without excessively increasing the computational cost.

The three distinct approaches for the mesh serve to illustrate the differences that can be obtained in numerical analysis by an increase in geometric fidelity of the model. Since vaults derive their stiffness and strength from their curvature, a faithful representation of their geometry is expected to give more accurate results.

The analyses were executed using the DIANA FEA software package (TNO 2017).. The analysis was performed under assumptions of material and geometric linearity and isotropy. This is consistent with the absence of visible cracking in the vault extrados and intrados after removal of the formwork.

4.2 Loading and boundary conditions

The models were loaded by their self-weight with no additional dead or live loads added. This is consistent with the state during and immediately after construction, during which minimal live load was present (small number of moving people and light equipment), as well as with the projected use of the vault: the overlaying floor will rest on the surrounding masonry walls and not on the vault itself.

Boundary conditions were applied according to the horizontal translational restraint provided by the underlying masonry structural elements. Therefore, the ties were not explicitly considered in the model. Vertical restraint was provided at the four corners, where masonry pillars are located. Horizontal restraint was provided in the direction of the neighbouring large vault (left side of the studied vault shown in Figure 3).

4.3 Material properties

In the absence of detailed experimental investigation of all the materials used in the construction of the vault, the elastic parameters of the vault masonry are determined using a combination of experimental testing, empirical expressions and design standard guidelines.

The mean compressive strength of the mortar f_m at three weeks after casting was determined equal to $4.20 \ N/mm^2$. The compressive strength of the solid clay units f_b is estimated at $8.00 \ N/mm^2$. From these values for the compressive strength of the two material components, the compressive strength of masonry f_k may be determined from the Eurocode 6 expression (CEN 2005):

$$f_k = K \cdot f_b^a \cdot f_m^b \qquad (1)$$

For the general purpose mortar, unit type and joint thickness used in the vault, the parameter K is equal to 0.55, a is equal to 0.70 and b is equal to 0.30. This results in a compressive strength for the masonry equal to $3.63 \ N/mm^2$ at three weeks after casting.

The Young's modulus of the masonry is calculated empirically. Drawing from a large number of experimentally derived ratios between Young's modulus E and compressive strength of masonry, the ratio may be determined from the expression (Drougkas, Roca, & Molins 2015):

$$E = 350 \cdot f_k \qquad (2)$$

This yields a Young's modulus of masonry E equal to 1270 N/mm^2 at three weeks after construction. Considering the compressive strength of mortar at two weeks after construction (3.38 N/mm^2) the resulting compressive strength of masonry is 3.40 N/mm^2 and the Young's modulus is equal to 1190 N/mm^2. This means that according to these expressions the increase in both these properties is roughly equal to 6.7% for an increase in the compressive strength of mortar equal to roughly 24.3%. While this increase is rather small, lime mortars continue to set for years, which can lead to further increase in the stiffness, if not offset by masonry creep or unforeseen extra loading.

Finally, the specific weight γ of the brick masonry is considered equal to 18 kN/m^3 and the Poisson's ratio ν is taken equal to 0.20.

The material properties of the masonry at three weeks after construction are summarized in Table 3. These are the properties used in the numerical analysis of the deflection after the removal of the formwork for all considered models, corresponding to three weeks after construction of the vault.

4.4 Analysis results

The results of the numerical analyses are summarized in Table 4, in terms of computed weight and maximum deflection at the centre of the vault. The shell and solid models give similar results in terms

Table 3. Mechanical properties of vault masonry at three weeks after construction.

Compressive strength	f_k	3.63	N/mm^2
Young's modulus	E	1270	N/mm^2
Poisson's ratio	ν	0.20	–
Specific weight	γ	18	kN/m^3

Table 4. Results of finite element analysis of masonry vault. Weight and central deflection of models.

Model	Weight [kN]	Central deflection [mm]
Shell	314	59.3
Solid	307	60.7
Scan-derived	352	32.2

of both parameters. The computed deflection, however, is nearly double of what was registered through manual and laser scanning measurements. The scan-derived model, conversely, has the highest weight (roughly 14.7 % increase compared to the shell model, mostly due to the topside fill), but results in the most accurate prediction of the maximum deflection. Its accuracy lies within the precision limits, equal to 2 mm, of the laser scanner.

The increased weight of the topside fill is offset by the added stiffness it provides in bending. Considering the increased weight in the solid and shell models would further increase the obtained deflection without an increase in the bending stiffness. The passing of the thrust line of arches and vaults through the fill near the supports is a regularly noted phenomenon that serves to stabilize such structures under vertical loading (Heyman 1966). Therefore, the inclusion of this fill has proven to be critical for the correct analysis of the vault.

All the models give similar deflection shapes, which are in good agreement with the deflected shape registered using laser scanning. These shapes are illustrated in Figure 10. This indicates the validity of the assumptions on the distribution of the load and, especially, the boundary conditions. Despite the difference in the number of degrees of freedom, the material and geometric linearity assumptions result in very short computation times for all models, including the scan-derived mesh.

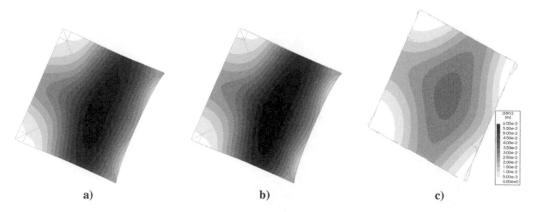

a)   b)   c)

Figure 10. Deflected shapes and deformation contours in [m] from finite element analysis: a) shell model, b) solid model and c) scan-derived model.

5 CONCLUSIONS

Simple damage monitoring techniques were shown to be used efficiently during reconstruction of a masonry vault. Without the use of complex mounting of equipment and the use of data acquisition systems, high accuracy deformation measurements are obtained with zero intrusion on the fabric of the building and no obstruction to the construction works.

Structural analysis of the vault has been carried out employing three approaches for representing its geometry. Large differences between the approaches in terms of obtained deflection highlight the importance of detailed geometric survey for the analysis of historic structures. This is true not only for the accurate determination of the self-weight load, but also for the correct estimation of the stiffness of the structure.

Detailed geometric survey data is shown to be critical in achieving accurate analysis results in structures whose behaviour is governed by their geometry. Vault structures are characterized by relatively complex geometry and derive their strength and stiffness from their geometric form. Therefore, laser scanning is proven to be critical for the correct application of structural analysis methods for such typologies.

ACKNOWLEDGEMENTS

The authors would like to thank V. Wirix from Denys NV and F. Noë from VK Engineering for supporting all on-site visits, and WTA-NL-VL for the financial support. Our thanks is extended to Stan Vincke, PhD candidate at the Geomatics Research group for assisting with the scanning of the vault.

REFERENCES

CEN (2004). *EN 1992- 1-1- Eurocode 2: Design of concrete structures - Part 1-1: General rules and rules for buildings*.

CEN (2005). *EN 1996- 1-1- Eurocode 6 - Design of masonry structures - Part 1-1: General rules for reinforced and unreinforced masonry structures*.

Como, M. (2012). *Statics of Historic Masonry Constructions*. Springer.

Drougkas, A., P. Roca, & C. Molins (2015). Numerical prediction of the behavior, strength and elasticity of masonry in compression. *Engineering Structures 90*, 15–28.

Geuzaine, C. & J.-F. Remacle (2009). Gmsh: A 3-D finite element mesh generator with built-in pre- and post-processing facilities. *International Journal for Numerical Methods in Engineering 79*, 1309–1331.

Girardeau-Montaut, D. (2015). Cloud Compareâ€"3d point cloud and mesh processing software.

Heyman, J. (1966). The stone skeleton. *International Journal of Solids and Structures 2*(2), 249–279.

ISCARSAH (2003). *Recommendations for the analysis, conservation and structural restoration of Architectural Heritage*.

Kazhdan, M., M. Bolitho, & H. Hoppe (2006). Poisson Surface Reconstruction. In *Eurographics Symposium on Geometry Processing*.

Milani, E., G. Milani, & A. Tralli (2008). Limit analysis of masonry vaults by means of curved shell finite elements and homogenization. *International Journal of Solids and Structures 45*(20), 5258–5288.

TNO (2017). DIANA Finite Element Analysis, User's Manual.

Verstrynge, E., L. Schueremans, & P. Smars (2012). Controlled Intervention: Monitoring the Dismantlement and Reconstruction of the Flying Buttresses of Two Gothic Churches. *International Journal of Architectural Heritage 6*, 689–708.

Preventive Conservation - From Climate and Damage Monitoring to a Systemic and Integrated Approach – Vandesande, Verstrynge & Van Balen (eds)
© *2020 Taylor & Francis Group, London, ISBN 978-0-367-43548-6*

Contribution of photogrammetry and sensor networks to the energy diagnosis of occupied historical buildings

S. Dubois, J. Desarnaud, Y. Vanhellemont & M. de Bouw
Belgian Building Research Institute, Limelette, Belgium

D. Stiernon & S. Trachte
Architecture & Climat, Université Catholique de Louvain, Louvain-La-Neuve, Belgium

ABSTRACT: The sustainable energy renovation of historical buildings is a challenge for all European countries. It is crucial for their conservation as well as for urban and rural development. Nonetheless, proposing adequate interventions requires appropriate investigation efforts. This paper presents a specific approach for performing the energy diagnosis of occupied historical buildings, developed under the constraints of several ongoing research projects. It is shown how photogrammetry and wireless sensor networks can be combined to produce rich datasets, while keeping disturbances for occupants at a minimum. Within this multi-disciplinary investigation program, the focus is also put on the production of input and validation data for implementing dynamic energy simulations. A case study is presented to illustrate the deployment of the proposed methodology. Only two site visits allowed to capture a large quantity of descriptive and performance information, which was valorized through clear protocols for sampling and formatting the data.

1 INTRODUCTION

1.1 *Energy retrofits of historical buildings*

In Europe, several research projects have recently focused on the issue of energy retrofitting of historical buildings and the identification of adequate intervention strategies (Martínez-Molina et al., 2016). Conserving built heritage does not mean 'freezing' it because that would make it unsuitable for present and future needs, in terms of use as well as from the perspective of comfort and performance (Fabbri, 2013). On the other hand, modifying ancient buildings to improve their energy performance cannot happen without questioning the application of standard solutions for energy optimization.

In Flanders, the project *ErfgoedEnergieLoket* (de Bouw et al., 2014) assists the architects specialized in architectural heritage regarding energy and comfort aspects. In Wallonia, the research project *P-Renewal* (Stiernon et al., 2017) aims to develop strategies for the sustainable retrofit of historical Walloon dwellings with heritage value and built before 1914. Within this latter project, whole-building dynamic energy models are used as flexible and exploratory tools to evaluate the relevance of different energy-related interventions. Nevertheless, calibrating building energy simulation (BES) tools is challenging, given the complexity of the underlying mathematical representations (O'Neill and Eisenhower, 2013) and uncertainty about the values of the input parameters.

1.2 *The building performance investigation in the 'digital era'*

In terms of energy efficiency and hygrothermal balances, it is well-known that traditional masonry buildings are specific: the indoor conditions are shaped by the high thermal inertia of walls and specific ventilation/infiltration patterns; their construction materials are the seat of complex coupled HAM (Heat Air and Moisture) transfers, which are not simple to model; the condition of these materials can influence the theoretical performance; the presence of building systems is often limited or outdated. Even if the final energy consumption is a central performance indicator in most building energy policies, this data alone is insufficient to grasp the building specificities and thus to propose adequate intervention strategies. The risk of accentuating existing pathologies or creating new ones is real. The biodeterioration of wooden beams subsequent to the internal insulation of a masonry wall is a classic example (Guizzardi et al., 2015).

For each retrofitting project involving a historical building, one crucial aspect is thus the implementation of a relevant multi-disciplinary investigation program, a hard and time-consuming task. It is aimed towards the characterization of the building 'as it is' and 'as it performs' through the mobilization of a range of sensing methods. Not only deficiencies and their causes are sought for, but also intrinsic qualities (*e.g.* natural comfort in some rooms) and values (*e.g.* building elements with heritage significance). The ultimate objective is to obtain organized information to judge the risks and benefits associated with a particular renovation measure, or a combination of measures. Studies on heritage buildings have encouraged the development of non-invasive technologies for such documentation and analyses (Troi et al., 2015).

A modern and systematic methodology was developed at the BBRI (Belgian Building Research Institute) to optimize the quality of the data collected in historical and occupied buildings, for exhaustive energy studies. It is based on state-of-the-art methodological approaches (Casanovas, 2008; Vieites et al., 2015) while integrating relevant and innovative digital tools and maximizing their potential of data production. Two non-destructive techniques (NDTs) are here used in a complementary way: multi-view photogrammetry and wireless sensor networks. The integration of those two key technologies into a methodological process was encouraged by specific constraints met during the *P-Renewal* project. This paper shows how they greatly benefit to the expert team to evaluate a building in terms of energy and comfort, with a view to its renovation.

2 UPDATING THE ENERGY DIAGNOSIS APPROACH

2.1 *Facing specific constraints during research efforts*

Within the *P-Renewal* project, five buildings from pre-1914 were studied, each of them being representative of one major type of dwellings in Wallonia (Stiernon et al., 2017). It was necessary to obtain a holistic picture of each one of them, in terms of their environment, composition, condition, occupancy and performance, while keeping the cost of tests to a minimum. For several reasons, leading in-depth energy studies in historical and occupied buildings were particularly relevant but also very challenging.

As the evaluation of future intervention strategies required properly calibrated whole-building energy models, the task was even harder. Indeed, many input parameters were required, several of whom are, unfortunately, very case specific. Each material present in the building has its own specificities, which must be translated into standard parameter values in the model. In addition, some physical variables should be monitored, ideally over a long period of time to provide dynamic input (e.g. climatic boundary conditions) or validation data to the model. The monitoring equipment can be invasive, and the installation cost significative.

An exhaustive evaluation of heritage values was another important step in this project. Even if the considered buildings were not specifically listed, the goal was to pinpoint features of significance across the investigated areas and estimate their state of preservation. Such analyses require a proper documentation phase that traditionally goes through meticulous on-site observations.

All studied buildings being permanently occupied, any technique that would allow to drastically speed up the on-site investigation and limit the impact on the occupants' activities would be precious. A clear diagnostic methodology had to be developed, with an optimization of the 'efforts-to-results' ratio and a limitation of the invasiveness of tests.

2.2 *Which key data for a comprehensive energy diagnosis of historical buildings oriented towards dynamic modelling?*

The energy diagnosis approach is first defined by the key information to be collected, with respect to the objective and the extend of the analysis pursued. Here, the research is characterized by a high exhaustivity. The target information is divided into four main domains, as illustrated on Figure 1.

The first data category groups the information related to the history of the building, its place in the past and present context, its evolution from its creation, and cultural values associated to its different components. Then comes the anthropological domain, which is a complex matter. It covers the data related to the behavior of occupants and their perception of this behavior. Those are crucial aspects to interpret any indoor air measurement or energy consumption information. The way the people occupy the different indoor spaces or the way they use and program the available heating systems during winter are some examples of information that can hardly be guessed from direct observation. Indeed, they are related to complex social and comfort schemes and can vary according to daily, weekly or seasonal pattern. The second anthropological aspect is related to the significance of the building and its components for the occupants, which, more than likely, will vary from the opinion of the heritage expert.

For reasons of convenience, technical data can be organized according to the study scale, which ranges from the material scale up to the site scale, the HVAC systems forming a specific category. Then, at each scale, the target data are classified either in the architectural/geometric domain or in the physics/performance domain. The first category groups descriptive information, such has the type and function of the identified building elements. The second category forms a wider domain where the ongoing hygrothermal phenomena are translated into condition-oriented aspects (e.g. a damage on the surface of a material) and performance-oriented aspects (e.g. the U-value of an envelope element).

As shown on the synthetic figure, part of the investigation data can be translated into input parameters for dynamic energy models. Each instance of a hygrothermal model requires a series of inputs: the geometry of the modelled region(s), the

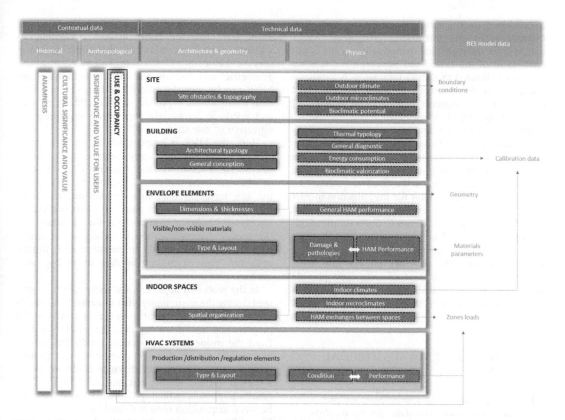

Figure 1. Data to be collected during the energy diagnosis of a historical building, with possible use as BES model input. Framed in solid black line, the aspects where 3D photogrammetry can facilitate on-site data collection; framed in dashed line, the aspects where wireless sensor networks provide a way of improving the investigation process. In grey letters, the contextual data that is often set aside from energy studies on existing buildings but are essential to propose heritage-compatible interventions through an adequate value assessment.

hygrothermal parameters of each material forming those regions, the hygrothermal conditions at their boundaries, and zone loads for BES simulations (e.g. occupancy conditions, heat sources). The impact of renovation or restoration interventions can only be evaluated once the model is properly calibrated. Such models are highly demanding and the confidence in simulation results is dependent on the quality and completeness of data.

2.3 Collecting and processing data

Once the key target data is identified, the appropriate diagnosis tools must be selected and their deployment organized, considering the occupation of the building and constraints in terms of access. Operationally, the energy diagnosis approach is divided into several phases.

First, the preliminary study, which aims to provide a general 'picture' of the building as it is and compile the first hypotheses about its composition and performance. It consists in a general

documentary investigation completed by one single preliminary visit, where the expert should try to optimize the quantity of collected data. As existing plans and photographs are often lacking, or only provide an incomplete image of the building, solely on-site observations will generate reliable information about its geometry. This geometric documentation is a crucial upstream step, as it will provide the basis to provide a context to any other information. The first visit is also an opportunity to briefly analyze the condition of the building, possibly by using handy non-invasive methods. Moreover, meeting the tenants of the building is always an extremely rich source of information. From the preliminary visit, the initial descriptions and hypotheses regarding the current hygrothermal behavior are compiled in an initial synthetic report.

In a second stage, a multidisciplinary testing program is deployed to enhance the information that was gathered during the preliminary study. This 'in-depth studies' phase also focuses on the validation of initial composition and performance hypotheses.

It combines on-site testing (*e.g.* blower door test, thermography) and monitoring campaigns, which extend over different time periods. Given the occupied status of the considered buildings, the number of visits is limited to a strict minimum. A synthesis phase closes the investigation where a final report compiles all the collected data and is communicated to the energy simulation specialist.

At those different stages, the expert relies on different investigation and communication tools. The more precise the description of a building must be, the more building visits are typically needed. There is thus a conflict between the need for exhaustive studies and the will to limit the investigation impact on occupants. On-site experts would thus benefit from any technique that allows to optimize the quantity of data collected in a given period. In the next sections, it is shown how photogrammetry and wireless sensor networks perfectly answer this need. However, the large amount of data they provide must be properly handled, transformed and synthetized, otherwise there is a risk of missing the initial objectives.

3 INTEGRATING INNOVATIVE DIGITAL TOOLS

3.1 *Photogrammetry as a multi-purpose and multi-scale tool to produce descriptive data*

When it comes to geometry, recent high-definition technologies have revolutionized the building surveying and recording processes (Guarnieri et al., 2006; Remondino, 2011; Yastikli, 2007), which are crucial when working on heritage. The documentation process is now benefiting from an extremely high level of details offered by such automatic 3D digitalization technologies. Among the available methods, 'Multiview Photogrammetry' (MVP) (Furukawa et al., 2015) is very promising as a multiscale and multi-purpose tool, not only for descriptive analysis but also for performance estimation. As its name suggests, the technique is based on the automatic processing of photographs in a software: the three-dimensional shape of an object is estimated from overlapping pictures with varying points of view. The recording process of heritage building studies can be greatly enriched by the level of details that MVP offers for highly textured objects.

From the raw 3D data that is produced from photographs, i.e. point clouds or triangular meshes, many useful deliverables can be produced. It includes 1D deliverables (e.g. roughness profile of the point cloud along a line), 2D deliverables (e.g. vectorized cross sections, 2D CAD drawings, screen captures, orthomosaïcs), 2.5D deliverables (e.g. planarity maps, distance maps, *cloud-vs-cloud* maps), 3D deliverables (e.g. textured meshes, cross sections of point clouds, 3D CAD drawings, annotated 3D PDF's, BIM

models), or simple figures or statistics. Deliverables can also be differentiated according to whether they focus more on geometric information or color information. In summary, the raw geometric and radiometric data stemming from MVP can be used for many building studies. This data transformation process is theoretically infinite and needs to be strictly supervised regarding the objective of the study.

Here, the MVP is proposed as a 'multi-purpose' tool for supporting the energy diagnosis of the building. So far, it seems that the potential of the technology has not been sufficiently exploited. In Figure 1, the key technical data that can be inferred from MVP through appropriate processing and analysis in clearly highlighted. Nonetheless, the difficulty for the energy diagnosis expert is rightly to determine how to transform the 3D information and, further upstream, how to collect the data in an appropriate way.

In the proposed approach, the technique is mobilized during the preliminary visit to capture the building completely and benefit from the data early in the project. The field team captures the whole building from the inside and the outside, producing two or more point clouds that are registered with ground control points. The capture distance is chosen to get a typical 'ground sample distance' of 2.5mm on the resulting photos, which provides a good balance between acquisition time and final model resolution. The different resulting point clouds are then processed to end up with the set of standard deliverables as illustrated in Table 1.

A first category of deliverables includes files that are directly generated from the reconstructed point cloud data and are called 'intermediate' deliverables. From those, secondary deliverables are created and shared to the involve expert team within the project. Those final files are all images, in order to facilitate the collaboration and ensure and optimal access to the desired data. In the future, the development of BIM (Building Information Modelling) should encourage the direct exchange of 3D information.

3.2 *Wireless sensor networks as a key facility to grasp the dynamic phenomena*

A holistic hygrothermal study of a building often implies the monitoring of physical variables. Whatever the quality of a photogrammetric survey, it will only give an 'instant' image of the building. The specificities of the hygrothermal behavior of an old building are profoundly dynamic: inertia and thermal comfort, for example, cannot be evaluated in a static way.

The sensors and systems dedicated to the monitoring phase are evolving constantly. Traditionally, research-oriented monitoring systems consisted of specialized data-logging stations connected to various wired sensors. Battery-operated sensors with embedded logging capabilities naturally succeeded

Table 1. A standard set of deliverables produced from the photogrammetric study and adapted for a holistic energy diagnosis. The data is ultimately processed in the form of images to facilitate the collaboration between the involved experts. See Figure 3 for a concrete example from a case study.

3D reconstruction data	Processed data (intermediate)	Processed data (final)	Use
Point cloud of indoor spaces (cleaned and registered)	→ I1. Distance map(s) of selected floor(s) and/or ceiling(s) compared to reference planar surfaces	→ I1-1. Orthoviews of the distance map(s)	*Evaluating the condition of floors and ceilings*
Point cloud of the envelope (cleaned and registered)	→ E1. Distance map of each façade compared to reference planar surfaces	→ E1-1. Orthoviews of the distance maps	*Evaluating the condition of walls*
	→ E2. Textured mesh of the envelope	→ E2-1. Orthomosaïc photos of all façades	*Materials/pathologies identification and mapping (through image analysis and machine learning)*
Merged point cloud of the envelope and the interior spaces (cleaned and registered)	→ M1. Cross-sectioned point cloud	→ M1-1. Isometric views of the sectioned model	*Illustrating the internal organization of the building*
	→ M2. 0.1m thick cross sections every 0.5m along main building axes	→ M2-1. Orthoviews of the sections	*Encoding the building geometry and the thickness of envelope elements in whole-building energy models*

with the development of low-power integrated circuits. Today the 'Internet of Things' is gaining popularity (Atzori et al., 2010) and many innovative wireless communication protocols are being deployed, allowing data to be transmitted remotely using radio frequencies (Gubbi et al., 2013). As a result, 'Wireless Sensor Networks' (WSN) were developed. In their simplest form, they combine sensor nodes and gateways for the end-transmission of data to the user (Bhattacharyya et al., 2010; Mottola and Picco, 2011). With WSN, all sensor measurements are more easily accessible because they are gathered in a single location in the building, and even stored on cloud servers. Parallel to the diversification of sensor communication schemes, the development of hardware and software based on an open-source approach has gained much attention (Fisher et al., 2015). The success of open-source development boards and the dynamism of user communities should encourage heritage experts to develop more WSN solutions tailored for heritage building studies.

During the multidisciplinary testing phase of the proposed methodology, wireless sensor networks (WSN) are valorized as flexible tools for remote data collection over long periods of time and for multiple buildings. Such networks can be easily deployed on many parallel sites and open-source microcontrollers allow custom-made solutions to be developed. For example, carbon dioxide, temperature and humidity can be monitored in multiple rooms, and heat fluxes on multiple surfaces. As summarized on Figure 1, WSN can improve the assessment of physical phenomena, energy consumption and building use.

4 CASE STUDY

4.1 The building

A traditional farm located in Enghien (Wallonia) is proposed here as a brief illustration to the systematic energy diagnosis approach (Figure 2). This XVIIIth century building was studied within the *P-Renewal* project. It is now used as a single-family dwelling, permanently occupied by three people. Several retrofit interventions have already taken place, including the replacement of some windows, the insulation of part of the attic floor and the installation of a thermostatic control of the central oil heater.

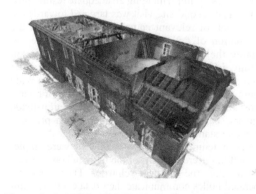

Figure 2. A traditional farm where the energy diagnosis approach was tested.

4.2 Preliminary visit and MVP study

After studying the available building documentation, which did not include any plans, and initializing contacts with the owners, the team prepared the first site visit. Once there, a MVP survey was led, both from the inside and from the outside of the building, where 857 and 111 photos were taken respectively. All photos were captured within a 3-hour time period with a Canon EOS 5D mark III equipped with a fixed 20mm Sigma lens and mounted on a tripod. The chosen diaphragm aperture was f/8, and the iso value of the sensor was set to a minimum of 100.

The resulting pictures were processed in *Agisoft Metashape*, ultimately forming two dense point clouds: 46 million points for the exterior photoset, 520 million for the interior photoset. All the rooms in the house were rebuilt without any significant alignment problems. Walls with a very uniform color, especially on the first floor, were reconstructed with less precision than the most 'textured' walls (such as the exposed brick walls in the attic). Six ground control points allowed the 'rigid' registration of both point clouds, with an average reconstruction error of 1.3cm on those control points.

Parallel to the photogrammetric survey, a brief humidity diagnosis was performed, using a protimeter to localize problematic zones within the envelope. Samples of salt efflorescences were also collected as part of the traditional prediagnosis assessments. Finally, the owner agreed to answer a standard interview, which provided basic information regarding the use of the building, the felt thermal comfort and observable pathologies.

From the collected information, the team responsible for the energy diagnosis could synthetize initial hypotheses regarding the building condition, its energy performance and the indoor comfort.

4.3 Second visit and installation of the WSN

As already mentioned, the in-depth studies focus on validating the initial composition and performance hypotheses by implementing an adequate testing program. The second site visit was targeted towards the realization of relevant complementary tests and the installation of a WSN. The latter was designed to capture the actual long-term hygrothermal variation within different rooms of the building. The used *Monnit* sensor nodes allow the monitoring of a wide range of 'standard' variables, such as the air temperature, but also the development of tailor-made sensors (based on a serial communication protocol bridge or standard voltage/current nodes).

Seven temperature/humidity sensors were implemented in different rooms of the farm and one outside to monitor the outdoor climate. Those battery-operated nodes communicate their data every 15 minutes to a gateway using 868 MHz radio frequency. Upstream of the network, a 3G router ensures that the

gateway can communicate with internet and reach the cloud servers where all data is stored.

The complementary performance tests consisted first in a blower-door evaluation of air infiltration, with varying control volumes to estimate air transfer schemes between parts of the building. Thermographic images were also taken from inside and outside of the building to highlight local air infiltration patterns or envelope defects. The humidity diagnosis was continued with some calcium carbide tests.

4.4 Valorization of MVP and WSN data

All the deliverables defined in Table 1 were generated from the photogrammetric reconstructions of the farm, as shown on Figure 3. As expected, those output files made it possible to address many aspects of the energy diagnosis. On the architectural level, the geometry of the building was adequately transcribed which allowed to properly contextualize all observations and tests. Using the 10cm thick cross sections through the whole building point cloud, not only floor plans could be inferred but also the thickness of all envelope elements. The connections of rooms and constructive elements appeared clearly, and many hypotheses could be put forward.

The great additional benefit of using MVP is the quality of color rendering it offers. In this matter, orthomosaïc photos were precious to complement the cross sections information. They served, for example, as basis for the listing of openings size and the mapping of pathologies.

In summary, the visual quality of the 3D point clouds, their geometrical resolution, and the overall geometric accuracy provided a strong base for conducting analyses and for presenting the results. For decision-making purposes, such data can be easily valorized in a BES software. In this study, *Openstudio* was specifically used to test various retrofitting strategies. The geometry of the model was created in *Sketchup* using a combination of cross sections images and orthomosaïc photos of the façades.

On the other hand, the WSN installation allowed to properly capture the inside and outside climatic conditions. The main benefit for the researchers was the permanent access to data. There was no need any more to plan building visits only to collect the sensor measurements. However, the multiplication of sensors and monitored sites led to a profound transformation of how the data is processed. For diagnostic purposes, each data set was traditionally analyzed manually, and graphs of interest were generated based on the observations. This time-consuming task is incompatible with the big data era. The development of specific *Python* scripts has thus been initialized to format the rich monitoring data offered by WSN. Those scripts focus on an automatic creation of relevant graphs and statistics for the studied period (Figure 4).

The WSN data was also transformed for incorporation as input data in the BES model. Again, *Python*

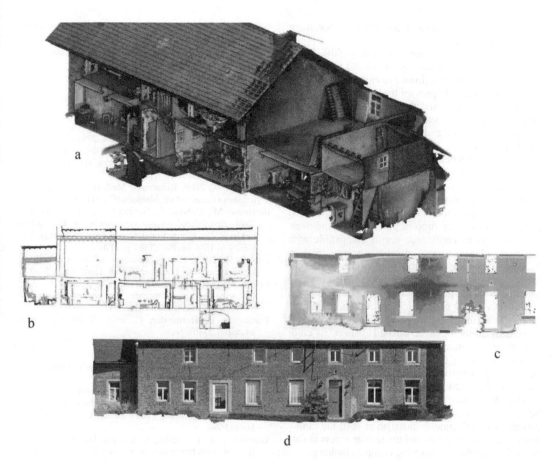

Figure 3. Some of the standard deliverables used for the energy diagnosis and modelling, generated with MVP for the reference case study. (a) Deliverable M1-1: isometric view of the merged point cloud with key cut sections; (b) Deliverable M2-1: orthoview of 10 cm thick cross sections along the X axis; (c) Deliverable E1-1: orthoview of a planarity map of the main façade; (d) Deliverable E2-1: orthomosaïc photo of the front façade, generated from a textured mesh.

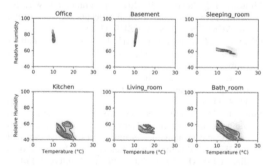

Figure 4. Automated charts creation: here, scatter plots are generated for 6 of the monitored rooms for one typical winter week.

scripting allowed to automatize the process. For example, the climatic boundary conditions were constituted as a combination of WSN on-site measurement and satellite data for radiation, wind and atmospheric pressure.

5 DISCUSSION

Digital technologies have enormous potential to improve the study of old buildings. Today, it is necessary to develop new diagnostic methodologies that fully incorporate them, with clear protocols for data transformation, valorization and archiving. Indeed, the digital era also means the data era. A bad definition of data handling could mean, at the end, no use of this incredibly rich data at all.

The case study showed that precise and very high-density 3D surveys can be created only from photosets. It offers great opportunities for the energy diagnosis. First, the method is nondestructive (remote sensing) while providing large possibilities in terms of analysis and reducing disturbances for occupants when working in 'real life' case studies. The method is also multi-scale. Depending on the type of photographic lens used and the typical capture distance, objects ranging from the microscopic up to the terrain scale can be digital-ized. Because the method is UAV-compatible, large, inaccessible or dangerous areas are also becoming easily diagnoseable.

MVP has however some noticeable pitfalls. Many factors can affect the quality of the 3D reconstruction and the protocol followed to capture the interest object has a major impact on the results. If the computational principles and the inherent limits are not properly understood, there is a risk of creating erroneous or incomplete data. Nonetheless, historical buildings that have undergone little transformation are good candidates for MVP campaigns. In this study, the use of photogrammetry alone on several case studies has proven that an absolute geometric error in the centimeter range is achievable. However, such mean value should not hide the fact the method can show some local drop in precision. On the overall, this level of confidence seems compatible with the required accuracy of dynamic energy models. Modern materials, with the predominance of synthetic and texture-less surfaces, are somewhat unsuitable for the proposed approach. There, when it comes only to 3D geometrical restitution, the terrestrial laser scanner is more reliable as it is less dependent on the operator skills. However, its color restitution is much less advanced than MVP. A combination of both techniques would bring the best to any study but is not always realistic from a cost point of view.

The WSN must be considered as an innovative infrastructure to collect dynamic data over long periods. The reduction of installation cost, the diminution of cable constraints and the remote access to data make them perfect for studying occupied buildings.

6 CONCLUSION

The proposed diagnostic methodology has already been successfully applied on several Belgian case studies, confirming clear benefits in terms of efficiency for the building energy diagnosis, especially when dynamic energy models are implemented to explore retrofitting scenarios and support the decision making. For the presented case study, it allowed producing a large quantity of data, from only two actual visits and minimum impacts to the occupants.

At the root of this approach, photogrammetry quickens up the reproduction of the building geometry, with a high level of detail. The combination of inside and outside surveys has proven to provide precious information regarding the architecture and performance of the building, and ultimately for implementing BES models. Not only the envelope geometry is easily transposable but cross sections through the 3D point clouds allows to infer the configuration of indoor spaces and the thickness of walls. Here, the exploitation of the colour information also plays a central role. It provides a strong basis for qualitative analysis (*e.g.* material identification, pathology diagnosis). On their side, wireless sensor networks facilitate and enlarge the collection of data relative to the dynamic behaviour of an occupied building. With an appropriate data processing scheme, both techniques are complementary and allow to rapidly implement accurate and exhaustive energy studies.

REFERENCES

Atzori, L., Iera, A., Morabito, G. 2010. The internet of things: A survey. Comput. Netw. 54, 2787–2805.

Bhattacharyya, D., Kim, T., Pal, S. 2010. A comparative study of wireless sensor networks and their routing protocols. Sensors 10, 10506–10523.

Casanovas, X. 2008. Rehabimed Method. Traditional Mediterranean Architecture. Montada (CAATEEB), Barcelona.

de Bouw, M., Dubois, S., Herinckx, S., Vanhellemont, Y. 2014. Specialized energy consultants for architectural heritage, in: International Conference in Energy Efficiency in Historic Buildings. p. 30th.

Fabbri, K. 2013. Energy incidence of historic building: Leaving no stone unturned. J. Cult. Herit. 14, e25–e27.

Fisher, R., Ledwaba, L., Hancke, G., Kruger, C. 2015. Open hardware: A role to play in wireless sensor networks? Sensors 15, 6818–6844.

Furukawa, Y., Hernández, C. 2015. Multi-view stereo: a tutorial. Delft: Now.

Guarnieri, A., Remondino, F., Vettore, A. 2006. Digital photogrammetry and TLS data fusion applied to Cultural Heritage 3D modeling. Int. Arch. Photogramm. Remote Sens. Spat. Inf. Sci. 36.

Gubbi, J., Buyya, R., Marusic, S., Palaniswami, M. 2013. Internet of Things (IoT): A vision, architectural elements, and future directions. Future Gener. Comput. Syst. 29, 1645–1660.

Guizzardi, M., Carmeliet, J., Dermoe, D. 2015. Risk analysis of biodeterioration of wooden beams embedded in internally insulated masonry walls. Cons. And Build. Mat. 99, 159–168.

Martínez-Molina, A., Tort-Ausina, I., Cho, S., Vivancos, J.-L. 2016. Energy efficiency and thermal comfort in historic buildings: A review. Renew. Sustain. Energy Rev. 61, 70–85.

Mottola, L., Picco, G.P. 2011. Programming wireless sensor networks: Fundamental concepts and state of the art. ACM Comput. Surv. CSUR 43, 19.

O'Neill, Z., Eisenhower, B. 2013. Leveraging the analysis of parametric uncertainty for building energy model calibration. Build. Simul. 6, 365–377.

Remondino, F. 2011. Heritage recording and 3D modeling with photogrammetry and 3D scanning. Remote Sens. 3, 1104–1138.

Stiernon, D., Trachte, S., de Bouw, M., Dubois, S., Vanhellemont, Y. 2017. Heritage value combined with energy and sustainable retrofit: representative types of old Walloon dwellings built before 1914. Energy Procedia, CISBAT 2017 International ConferenceFuture Buildings & Districts – Energy Efficiency from Nano to Urban Scale 122, 643–648.

Troi, A., Bastian, Z. 2015. Energy Efficiency Solutions for Historic Buildings. Berlin: De Gruyter.

Vieites, E., Vassileva, I., Arias, J.E. 2015. European Initiatives Towards Improving the Energy Efficiency in Existing and Historic Buildings. Energy Procedia 75, 1679–1685.

Yastikli, N. 2007. Documentation of cultural heritage using digital photogrammetry and laser scanning. J. Cult. Herit. 8, 423–427.

Preventive Conservation - From Climate and Damage Monitoring to a Systemic and Integrated Approach – Vandesande, Verstrynge & Van Balen (eds)
© 2020 Taylor & Francis Group, London, ISBN 978-0-367-43548-6

Author Index

Achig, C. 103

Barsallo, G. 19
Barthel, A. 125
Bassier, M. 137

Caldas, V. 19
Canali, F. 93
Cantini, L. 93
Cardoso, F. 19
Chen, Y. 87

de Bouw, M. 145
de Koning, S. 71
Della Torre, S. 11, 93
Desarnaud, J. 145
Drougkas, A. 137
Dubois, S. 145

Ferreira, T.C. 77

Gao, Y. 87
García, G. 103
Ge, Z. 87
Geyer, C. 119
Ghazi Wakili, K. 125

Gwilliam, J. 129

Ižvolt, P. 59

Konsta, A. 93

Li, E. 87
Li, J. 87
Linskaill, S. 67
Liu, B. 37

Meiping, W. 27
Moioli, R. 47
Müller, A. 119

Naldini, S. 71, 113

Peñaherrera, C. 19
Prizeman, O. 129

Rodas, T. 19
Rong, Q. 37

Shea, A. 129
Shi, H. 27
Stahl, Th. 125

Stiernon, D. 145

Tenesaca, P. 19
Tenze, A. 103
Tracht, D. 125
Trachte, S. 145

van de Grijp, E. 71
van de Varst, G. 71
van Hees, R.P.J. 113
Vandesande, A. 3
Vanhellemont, Y. 145
Vergauwen, M. 137
Verstrynge, E. 3, 137

Walker, P. 129
Wehle, B. 119
Whitman, C.J. 129

Xinjian, L 27
Xu, D. 87

Zhang, J. 37
Zhou, Y. 87

Printed and bound by CPI Group (UK) Ltd, Croydon, CR0 4YY

24/10/2024

01778295-0012